Economic growth &
employment structure

Economic growth & employment structure

A study of labour outmigration from agriculture in developing countries

A. K. Ghose

International Labour Office Geneva

First published 1990

Ghose, A.K.
– – *Economic growth and employment structure: A study of labour outmigration from agriculture in developing countries.* Geneva, International Labour Office, 1990
/WEP pub/. /Economic growth/, /Agricultural employment/, /Rural migration/, /Trend/s 1969-85, /Developing country/s. 13.01.3
ISBN: 92-2-107278-9

ILO Cataloguing in Publication Data

The designations employed in ILO publications, which are in conformity with United Nations practice, and the presentation of material therein do not imply the expression of any opinion whatsoever on the part of the International Labour Office concerning the legal status of any country, area or territory or of its authorities, or concerning the delimitation of its frontiers.
The responsibility for opinions expressed in signed articles, studies and other contributions rests solely with their authors, and publication does not constitute an endorsement by the International Labour Office of the opinions expressed in them.
Reference to names of firms and commercial products and processes does not imply their endorsement by the International Labour Office, and any failure to mention a particular firm, commercial product or process is not a sign of disapproval.

ILO publications can be obtained through major booksellers or ILO local offices in many countries, or direct from ILO Publications, International Labour Office, CH-1211 Geneva 22, Switzerland. A catalogue or list of new publications will be sent free of charge from the above address.

Printed in Switzerland VAU

Preface

This study was planned as part of the research programme of the Rural Employment Policies Branch of the International Labour Office. Its basic aim is to provide a global review of the issues relating to agricultural employment in developing countries. The point of departure is a recognition of the fact that problems of employment and poverty in agriculture are never resolved within the framework of the agrarian economy alone. Indeed, economic development is always associated with a steady decline in the relative weight of agricultural employment. The nature and evolution of the problems of employment and poverty in agriculture, therefore, have to be viewed in the context of overall development strategies.

The study assembles an impressive volume of statistical data covering 86 developing countries. It focuses on two major topics of interest, namely, the process of labour outmigration from agriculture and the trends in underemployment in agriculture. The major findings of the study, which covers the period 1969-85, can be summarised as follows:

(i) the pace of labour outmigration from agriculture has been similar to that in the early phases of industrialisation of today's developed countries;

(ii) in the Asia-Pacific region and sub-Saharan Africa, however, the rate of outmigration was not sufficient to stem the fast decline in the land/worker ratio, which was already low;

(iii) the rate of growth of employment in the modern non-agricultural sector generally exceeded that of the non-agricultural labour force in the seventies, but modern sector employment stagnated or declined in many countries;

(iv) the pace of labour outmigration from agriculture, however, was not much affected by the trends in modern sector employment;

(v) in the seventies, underemployment in agriculture declined in all regions except sub-Saharan Africa;

(vi) in the eighties, this declining trend was really sustained only in the Asia-Pacific region;

(vii) given the pace of labour outmigration from agriculture, the trends in underemployment in agriculture depend on the pace of land-augmenting technological change.

These findings, the author argues, cast doubt on many of the received theories. A closer examination shows that factors such as Engel's Law, changing comparative advantage in external trade, differential growth of sectoral productivities and rural-urban wage (or income) differentials do not explain the observed patterns of labour outmigration from agriculture in a satisfactory manner. The author, therefore, suggests some alternative hypotheses. He proposes a three-sector analytical scheme – the three sectors being agriculture, traditional non-agriculture (incorporating rural non-farm and urban informal sectors) and modern non-agriculture. "Push" factors determine a base rate of outmigration from agriculture into traditional non-agriculture; modern non-agriculture draws labour from traditional non-agriculture in accordance with the former's need. This need, moreover, depends on the rate of growth of modern manufacturing. On the basis of such a perspective, the author discusses some major issues of employment and macro-economic policy confronting the developing countries.

Like most studies of the type, this study raises more questions than it can answer, thereby suggesting areas of further research. It will obviously be desirable to seek confirmation of some of the findings through region- or country-specific research. But there are two other areas where considerable research efforts need to be devoted: *(a)* the relation between underemployment and poverty in agriculture, and *(b)* the characteristics of underemployment and poverty in traditional non-agriculture, i.e. rural non-farm and urban informal sectors. In all these areas, there is also an urgent need to improve the data base.

It is hoped that this study takes us a step further in understanding the classical issue of the role of agriculture in economic and social development.

S. Radwan
Chief, Rural Employment Policies Branch

Acknowledgements

The author* wishes to thank Albert Wagner whose computational assistance made the study possible. At an early stage of the work, the preliminary ideas and findings were presented at a World Employment Programme Research Seminar and the author greatly benefited from the comments and suggestions made by the participants. Thanks are also due to K. Griffin, S. Radwan and W. Van Ginneken who read through the penultimate draft and gave very useful comments. The author alone is responsible for the opinions and the errors, if any.

* Senior Research Economist, Rural Employment Policies Branch, Employment and Development Department, International Labour Office, Geneva.

Contents

Contents

List of tables

1

Introduction

1.1 The context

In development theory, the relation between economic growth and employment is conceived in terms of intersectoral shifts of the labour force and also of changes in the mode of employment. A developing economy, almost by definition, is primarily agrarian and is characterised by the predominance of self-employment and the existence of surplus labour. Economic growth is associated with the development of industries and services which are modern in two senses: they use capital (produced means of production) and they employ labour on the basis of a wage system. Consequently, economic growth is accompanied by a growth of regular (full-time) wage employment in non-agriculture and a corresponding decline of surplus labour in agriculture (which also means decline of self-employment). This movement of labour, moreover, is itself a major source of economic growth until a certain turning point is reached (when surplus labour disappears), since the workers who move have zero marginal productivity in agriculture but positive and usually high marginal productivity in modern non-agricultural sectors.

However, a large number of empirical studies have shown that the movement of labour from agriculture to industry and services is a characteristic of the process of economic growth in all countries. At one level, this finding has served to strengthen the view that industrialisation is the key to economic growth in developing countries so that rapid creation of wage employment is also the key to a solution of the employment problem. But at another level, the empirical regularities have generated an unfortunate tendency to suppose that the contemporary developing countries are what the developed countries were in the earlier phases of their history. Such a view led to the development of analytical descriptions of the processes of structural change which were thought to be of general validity rather than of particular relevance to developing countries. Some of the particularities of the developing countries tended to be analytically de-emphasised as a result.

In the relevant literature, for example, the intersectoral shifts and structural changes in employment which accompany growth are seen to result from autonomous changes in the structure of demand and in sectoral labour productivities. Changes in the structure of domestic consumer demand are accounted for by the celebrated Engel's Law. The changes in the structure of external demand for exports are accounted for by analogous factors operating at the level of the world economy. These tendencies are supposed to combine with rapid growth of labour productivity in agriculture to generate movements of labour from agriculture to industry and services.

Even at a superficial level, these explanations do not seem to fit the experience of developing countries very well. Given the persistent problem of mass poverty in developing countries, it is hard to believe that Engel's Law has been of much significance in these countries. While the share of agricultural products in merchandise exports has declined for the developing countries as a group, the decline has in fact been significant only for a few countries. And labour productivity in agriculture actually stagnated or declined in a large number of countries. Yet, labour outmigration from agriculture has been significant and universal.

A second example concerns the factors which motivate labour to move from one sector to another. In the standard analysis (which implicitly assumes full employment), as the structure of production responds to the changing structure of demand, changes in the pattern of demand for labour manifest themselves in terms of intersectoral wage differentials. A movement of labour then occurs from the relatively low wage sector to the relatively high wage sector. This idea, with suitable modifications, has also been used to interpret the process of migration of labour from agriculture into non-agriculture in developing countries. The very *differentia specificae* of a developing country – the existence of surplus labour and the predominance of self-employment in agriculture – are effectively lost sight of in the process.

None of these hypotheses is in fact needed to explain labour outmigration from agriculture in developing countries once the specificities of these countries are adequately recognised. In the dual economy theories, for example, the autonomous element is supposed to be capital accumulation in the modern non-agricultural sector. This sets in motion a growth process which involves continuous withdrawal of surplus workers from agriculture and their employment in modern non-agriculture at an institutionally given wage. The process is associated with a rising share of profits in national income and in so far as there is consumption out of profits, the share of food in total consumption expenditure may indeed decline. But this is an effect of distributional changes and not a reflection of Engel's Law; no significant change in the consumption pattern of the mass of population is involved. The export structure may change, but this can be seen as a consequence rather than as a cause of changes in the production structure. As for the motivation for labour to move out of agriculture, this can be understood in terms of attempts by agricultural households to maximise income.

Even these ideas, however, leave certain facts about the process of structural change in developing countries unexplained. A widely noted fact is that the so-called urban informal sector has emerged, in many developing countries, as a major reservoir of surplus labour and a major domain of self-employment. Clearly, labour did not always move from self-employment to wage-employment nor did it necessarily move from low-productivity jobs to high-productivity jobs.

A second fact, little noticed so far, is that even when economic growth was slow or negative, labour outmigration from agriculture still occurred. This is evident from the experience of Latin American and African countries in the eighties. In these countries, wage-employment in the modern sector often declined in absolute terms in the eighties, but the pace of labour outmigration from agriculture hardly changed. This is difficult to reconcile with the hypothesis that capital accumulation in modern non-agriculture is the *primum mobile* for structural change.

A way out of these puzzles is to hypothesise that intersectoral shifts of the labour force cannot be equated to structural change in the context of developing countries. In that case, the dual economy perspective can be regarded as adequate for understanding structural change but not for studying intersectoral shifts of labour force. But the hypothesis has several other analytical consequences. In the first place, intersectoral shifts of the labour force cannot be used as indicators of improvements or deterioration in employment conditions; the true indicator is the share of wage-employment in total employment in non-agriculture. Secondly, the issue of labour absorption in agriculture needs to be placed in a proper perspective. The argument that because the modern sector is unable to employ much of the incremental labour force, agriculture needs to maximise labour absorption loses much of its force, since it appears that the pace of labour outmigration from agriculture is unlikely to be much affected. This does not imply that maximisation of labour absorption in agriculture is not a desirable objective. But it does suggest that this cannot serve as a substitute for rapid employment creation in the modern sector. This means, thirdly, that the problem of capital accumulation in the modern sector deserves prime attention and agriculture's contribution to this needs to be carefully assessed. Agriculture's roles as a supplier of basic wage goods (food) and as a major source of demand for the modern sector's products are no less significant than its role as a provider of employment.

1.2 The objectives

The brief discussion above outlines the major issues which the present study seeks to explore. The study focuses exclusively on developing countries. It attempts to establish and explain the basic trends for the period 1969-1985 in two inter-related areas – outmigration of labour from agriculture and underemployment in agriculture – and goes on to propose a policy perspective. Issues of employment and underemployment in non-agriculture are discussed only to the extent that they help explain the trends under consideration. Since the period under scrutiny covers a period of prosperity (1969-79) and a period of severe economic difficulties (1979-85) due to global recession and debt problems, particular attention is paid to the effects of economic crisis. Indeed, as we shall see, a comparison of the results for the two sub-periods proves very useful for clarifying a number of analytical issues. It would have been desirable to extend the period up to 1988 but this was not possible because of lack of available data.

The study thus aims to be both informative and analytical. Indeed, it is perhaps worth stating that one of the motives for undertaking the study was the opportunity it gave to assemble the relevant statistical data, presently scattered in a variety of sources, and thus to build an information base for continual monitoring and analysis, at a global level, of the issues concerned. The annexed statistical tables, in addition to forming the basis of the global review and cross-country analysis undertaken in this study, should in themselves be of interest to all those concerned with the issues discussed here.

The reason for focusing exclusively on developing countries should be evident from the discussion in the preceding section. Although labour outmigration from agriculture is a common feature of all countries, the forces governing it in developing countries are very different from those in developed countries. And underemployment in agriculture is what distinguishes a developing from a developed country. A lumping together of developing and developed countries, therefore, is analytically unsound. Of course, the developing countries themselves are quite heterogeneous and it is arguable that the differences between Brazil and Portugal are less significant than those between Brazil and Burundi. The point, however, is to minimise the possibility of distortions; the fact that it cannot be completely removed does not invalidate this.

It will be noted that the dichotomy adopted here is agriculture/non-agriculture rather than rural/urban. This choice was dictated by the fact that data on employment and the labour force generally refer to major occupations or activities irrespective of geographical location. The definition of urban (and therefore of rural) areas, moreover, differs from country to country and changes over time in any given country so that the rural/urban dichotomy is probably less appropriate for empirical analysis in any case. It has to be noted, however, that both dichotomies have been used in the development literature without much concern for the differences between them. In growth analysis, the agriculture/non-agriculture dichotomy is typical while in the analysis of certain topics such as migration and poverty, the rural/urban dichotomy is commonly used. Such anomalies do create problems which we cannot hope to solve in this study. Fortunately, these problems are not so serious as to make coherent analysis impossible.

One final remark. The study is concerned with a selected set of issues and does not aim to be comprehensive in any sense. There are a number of issues, not unrelated to the topics discussed here, which are not discussed in any substantive manner. These include the effects of labour outmigration on the age- and sex-composition of the agricultural labour force, the significance of rural non-agricultural activities, open unemployment and certain aspects of the functioning of labour markets. On some of these issues, a good deal of research is needed before analytically interesting conclusions can be reached; on others, a good deal of research already exists and not much that is new can be said at this point.[1]

1.3 The methodology

The study bases itself on statistical data relating to 86 developing countries. A few of the developing countries had to be excluded because data on some crucial variables (e.g. GDP, labour force, etc.) were not available. Certain other developing

[1] The research on the effects of labour outmigration from agriculture on the age- and sex-composition of the agricultural labour force and their significance is still at early stages. The research on rural non-agricultural activities is more advanced and a good review is available in Saith (1989b). A good deal of research has been done on the functioning of rural labour markets. Cf. Binswanger and Rosenzweig (1981); Ray (1987); de janvry and Sadoulet (1987); Bharadwaj (1989); Rosenzweig (1988); Morawetz (1975) and Squire (1981).

countries were deliberately excluded either because they had a population of less than one million in 1969 or because they had an insignificant agricultural sector. It is to be noted that with respect to some of the variables used in the study, data were not available for all 86 countries. The number of countries for which data were available is indicated as appropriate throughout the study.

For descriptive purposes, the countries have been classified into four groups: sub-Saharan Africa (SSA), Asia-Pacific (AP), Latin America-Caribbean (LAC) and Near East-North Africa (NENA); the details are given in Appendix A. Classifications are always problematic and grouped data always incorporate some distortions. The possible alternative classification schemes (e.g. by GDP per capita, by export structure, etc.), however, were considered unsuitable for our purposes. Given the issues with which we are concerned here, it is desirable to ensure a certain level of uniformity of agrarian structure among the countries in a group. From this point of view, classification by region appears to be the best of the available alternatives. And regional perspectives, it should not be forgotten, are of interest in themselves. Nevertheless, it has to be acknowledged that with respect to certain variables, the average picture for a region may hide more than it reveals or may even be misleading. In such cases, we have tried to double-check by examining country-level evidence.

As a summary measure of variables for each group, we have chosen to use the median rather than the weighted average. One reason is that the weighted average for any variable would always be dominated by the values pertaining to large countries in a group and this can distort the average picture. But there are more practical reasons. Estimation of weighted averages usually involves adding up the values of a variable for the countries in a group. This requires all values of a variable to be expressed in the same unit. In many instances (e.g. population), this poses no problem; but in others (e.g. GDP) it involves problems of conversion. Moreover, the fact that the data set is incomplete (i.e. there are missing values) gives rise to another kind of problem. Consider, for example, a group consisting of 20 countries. If, for a particular variable, there are complete time series for 18 of the countries but shorter time series for the other two, then for estimating the weighted average growth rate, either two countries have to be dropped or having a shorter time series for the whole group has to be considered. Yet, for the two countries with a shorter time series, the growth rates need not be appreciably different from what they might have been had longer time series been available. The use of the median helps to avoid these types of problems.

Simple cross-country regressions have been employed for two kinds of purposes. In some cases, they have been used to study ex-post relationships among variables; the questions here are open-ended (e.g. did the countries with higher per capita incomes also have higher rates of growth of population?). In others, they have been employed to test hypotheses which are derived from *a priori* reasoning. In such cases, we have definite expectations about the nature of the relationship between variables.

This last point is worth emphasising. In a study such as this, it is not possible to ensure complete reliability of statistical data. Besides, there are conceptual difficulties involved in using the results derived from cross-country analysis for interpreting developments over time in particular countries or regions. For these reasons, we have chosen not to emphasise quantitative estimates of parameters. Rather, the objective has been to examine the extent to which the empirical evidence is consistent, in qualitative terms, with theoretical expectations in particular instances.

1.4 Organisation of the study

The study begins with an investigation, in Chapter 2, of the patterns of growth of population and labour force; this is a necessary preliminary though some findings of analytical interest also emerge. In Chapter 3 a factual assessment of the pace and pattern of labour outmigration from agriculture is first provided. The basic determinants of labour outmigration are then studied. Chapter 4 is concerned with underemployment in agriculture. Its relation to rural poverty is first discussed. This discussion leads on to an analysis of time-trends and their relation to the patterns of agricultural growth and labour outmigration from agriculture. In Chapter 5 the major conclusions, both factual and analytical, are summarised and an outlook for the future is deduced therefrom. Some major issues of macro-economic policy are then discussed.

2

Growth of population and labour force

The discussion in this chapter is intended to provide the necessary backdrop for the analysis which follows in the subsequent chapters. Most of the facts about the patterns of growth of population and labour force are well-documented; the weaknesses of the data are also quite well-known. The discussion here, therefore, neither makes any attempt to be factually comprehensive nor considers limitations of the data. Instead, it focuses on two particular aspects which are of some relevance for the present study. The first of these concerns the implications of the observed cross-country patterns for demographic transition in individual countries; this is useful for developing an outlook for the future. The second aspect concerns the relationship, at a cross-country level, between demographic expansion and economic growth. This is an analytically important issue and has been much discussed in the literature.

2.1 Some aspects of demographic growth

The developing countries are confronted with a challenge of employment creation which is quite unprecedented in history. And one of the reasons is that they have been experiencing rates of demographic expansion which are much higher than those experienced by today's industrialised countries at comparable levels of development (the other reason is that they start with a sizeable stock of surplus labour). All this is quite well-known[1] and the data in table 2.1 merely updates the facts. During 1969-85, the rate of growth of population in developing countries was well above 2.5 per cent per annum; this compares with a rate of 0.8 per cent in developed countries during 1800-1900. Roughly the same observations can be made about the rate of growth of labour force. In developing countries, moreover, both the rate of growth of population and that of labour force have been accelerating since 1900 and the process of acceleration continues, as the data in table 2.1 show.

A few other particular points are worth noting. First, viewed across regions, the rate of growth of population was highest in Near East-North Africa and lowest in Asia-Pacific. Secondly, there are clear signs of continuing acceleration in sub-Saharan Africa and of deceleration in Asia-Pacific; in Near East-North Africa and Latin America-Caribbean, the rate of growth of population appears to have been stable. Thirdly, the effects of acceleration or deceleration in the rate of growth of population did not really show up in the rate of growth of labour force during the period

[1] Cf. Bairoch (1975), Squire (1981), World Bank (1984), Kelley (1988) and Birdsall (1988).

Table 2.1. Growth of population and labour force (median values)

	Population		Labour force	
	1969-79	1979-85	1969-79	1979-85
SSA	2.66 (39)	2.87 (39)	2.12 (39)	2.19 (39)
AP	2.33 (15)	1.96 (15)	2.21 (15)	2.36 (15)
LAC	2.67 (21)	2.61 (21)	2.91 (21)	2.92 (21)
NENA	3.04 (11)	3.04 (11)	3.19 (11)	3.43 (11)
DC	2.53 (86)	2.68 (86)	2.52 (86)	2.56 (86)

Note. Figures in parenthesis indicate the number of countries for which data were available.
Source. Appendix table I.

considered; the latter showed signs of acceleration in all regions. The rate of growth of labour force, of course, necessarily has a lagged response to changes in the rate of growth of population and the indication is that the length of the lag is at least 15 years. This is an issue which we cannot explore here; but it is perhaps worth noting that since a major cause of the acceleration in the rate of growth of population in developing countries is the decline in infant mortality, a 15-year lag seems quite plausible.[2]

2.2 Demographic growth and level of development

One interesting fact emerges from the regression results presented in table 2.2: while, during the period under consideration, the rate of growth of population was unrelated to the level of development (as measured by the initial GDP per capita) across countries, the rate of growth of labour force did show a positive relation with the latter. Admittedly, the coefficient is rather small but it is significantly different from zero. While a full exploration of the factors underlying these features would be out of place here, a few relevant observations can be made.

One explanatory hypothesis we could consider is that the rate of labour force participation tended to increase with growth. This would imply, at a cross-country level, that countries with a higher GDP per capita tended to have a higher rate of labour force participation. This, however, was not the case. As can be seen from the results presented in table 2.3, the rate of labour force participation was actually inversely related with GDP per capita across countries. A little reflection shows that this is what we should expect. As economic growth proceeds, labour force participation of children and youth goes down because of the growing importance attached to education. Furthermore, because economic growth is associated with the growing importance of regular wage-employment, the labour force participation of older people also declines. In some situations, these tendencies may be counteracted to some extent by an increase in the labour force participation of women. But this is not universal, since cultural factors exert a major influence on female labour force

[2] For a discussion of the issues involved, see Bloom and Freeman (1987).

Table 2.2. Linear regression results

Period	Constant	Initial GDP per capita[a]	R^2	n
Dependent variable: Population growth rate				
1. 69-79	2.53	0.0003	0.005	83
	(21.18)	(0.62)		
2. 79-85	2.66	0.0002	0.004	86
	(22.79)	(0.61)		
Dependent variable: Labour force growth rate				
3. 69-79	2.25	0.0009*	0.10	83
	(17.21)	(2.98)		
4. 79-85	2.24	0.0008*	0.10	86
	(18.93)	(3.10)		

Notes. [a] In 1970 US dollars; refers to 1969 for regressions 1 and 3 and to 1979 for regressions 2 and 4. * Significant at 1 per cent. Figures in parenthesis are t-statistics. Log-linear regressions were tried but are not reported; they do not improve the results.

participation. On balance we should expect a *ceteris paribus* decline in the rate of labour force participation as economic growth occurs and, therefore, a negative relation between the rate of labour force participation and GDP per capita at a cross-country level.[3]

It thus appears that the positive relation between the rate of growth of labour force and the level of development is found not because of but rather in spite of the fact that the participation rate varies inversely with the level of development. The burden of explanation, therefore, must fall on the past pattern of population growth. Assuming a fifteen-year lag[4] between changes in population growth rate and the response of the labour force growth rate to these changes, the hypothesis can be stated as follows: for the fifteen-year period proceding 1969, the rate of population growth was positively related to the level of development across countries.

The implicit logic is that the countries which initiated modern economic growth (to use Kuznet's term) earlier also embarked on the path of demographic transition earlier.[5] At a cross-country level, this implies that the relation between the rate of population growth and the level of development would follow a particular sequence. It would be positive at first, then non-existent, then negative and then non-existent again. The relation beween the rate of growth of labour force and the level of development would follow the same sequence with a 15-year lag.

[3] For a detailed discussion of these aspects, see Standing (1978).

[4] This obviously is a rough approximation.

[5] Demographic transition refers to the transition from near equality of birth and death rates at high levels to near equality of birth and death rates at low levels. In general, the decline in death rates precedes that in birth rates. Consequently, the relation between the rate of growth of population and the level of development can be represented by an inverted U curve.

Table 2.3. Regression results, dependent variable: Labour force participation rate

Type	Year	Constant	Initial GDP per capita	R^2	n
Linear	1969	43.44 (31.47)	−0.01* (−4.37)	0.19	83
Linear	1979	42.83 (32.69)	−0.01* (−4.51)	0.19	86
Linear	1985	42.23 (31.20)	−0.01* (−3.51)	0.14	77
Log-Linear	1969	4.58 (28.97)	−0.18* (−6.05)	0.31	83
Log-Linear	1979	4.38 (30.03)	−0.14* (−5.26)	0.25	86
Log-Linear	1985	4.25 (29.37)	−0.11* (−4.32)	0.20	77

Notes. * Significant at 1 per cent. Figures in parenthesis are t-statistics. The participation rate is defined simply as the ratio of total labour force to total population.

What all this suggests is that over the next 15-year period, we should expect to see a negative relation between the rate of growth of population and the level of development and no relation between the rate of growth of labour force and the level of development. The rate of population growth will decelerate in countries with relatively high GDP per capita and will accelerate in countries with low GDP per capita. The rate of growth of labour force, too, will accelerate in low-income countries but will remain stable in high-income countries.

Translated into a regional perspective, the outlook appears to be as follows. The high-income countries of Asia-Pacific and Latin America-Caribbean will complete, by the end of the century, the penultimate stage of their demographic transition, i.e. the rate of population growth will stabilise at a low level and the rate of growth of the labour force will begin to decelerate. The low-income countries, mostly in sub-Saharan Africa, will only reach the second stage of demographic transition, i.e. the rate of population growth will stabilise at a high level and the rate of growth of the labour force will continue to accelerate. The middle-income countries, scattered around in Near East-North Africa, Asia-Pacific and Latin America-Caribbean, will reach the third stage of demographic transition, i.e. the rate of growth of population will begin to decelerate and the rate of growth of the labour force will stabilise at a high level. It can be said that, on balance, sub-Saharan Africa will experience, in the next 15 years to come, accelerating population and labour force growth; Asia-Pacific and Latin America-Caribbean will experience decelerating population growth and stable labour force growth; and Near East-North Africa will experience stable population growth and accelerating labour force growth.

These are, of course, generalisations which necessarily embody distortions. Some low-income countries (e.g. China and India), through the implementation of family planning programmes, have already reached the penultimate stage of their demographic transition. Vigorous implementation of family planning programmes in some countries in the immediate future will clearly also modify the picture. Nevertheless, all this is unlikely to alter the basic conclusion: the poorer countries will be facing a bigger challenge of employment creation than the richer countries.

2.3 Demographic change and economic growth

As the regression results in table 2.4 show, neither the rate of population growth nor the rate of growth of the labour force had any relationship with the rate of GDP growth. In fact, these results would not be worth reporting were it not the case that a relationship is often postulated in the development literature.

At initial stages, economic growth affects demographic change through reducing mortality; only at a later stage, when the quality rather than the size of the labour force becomes relevant, does economic growth induce decline in fertility. It might seem, therefore, that we should expect the rate of population growth to be positively related to that of GDP growth. However, a significant reduction in mortality usually occurs at the very start of the growth process as a result of control of some simple but widespread diseases (e.g. malaria, typhoid, etc.). After that, the decline in mortality is a slow process for a variety of reasons. In the first place, no monotonic relationship between economic growth and development of health services can be postulated. Some high-growth countries devote few resources to the development of health services while some low-growth countries devote substantial amounts. Secondly, as will be clear from the discussions in the subsequent chapters, economic growth in developing countries has not always been associated with a steady

Table 2.4. Linear regression results

Period	Constant	GDP growth rate	R^2	n
Dependent variable: Population growth rate				
1. 69-79	2.60	−0.002	0.00	86
	(17.96)	(−0.06)		
2. 79-85	2.60	−0.0003	0.00	85
	(29.01)	(−0.01)		
Dependent variable: Labour force growth rate				
3. 69-79	2.29	0.05	0.03	86
	(13.64)	1.55)		
4. 79-85	2.47	0.02	0.007	85
	(25.99)	(0.77)		

Note. Figures in parenthesis are t-statistics.

improvement in food consumption and the nutritional status of the population.[6] At a cross-country level, moreover, we are observing countries at different stages of demographic transition. When we put all the observations together, it becomes clear that we should not expect a systematic relationship between the rate of population growth and that of GDP growth.

It is an old argument, however, that a high rate of population growth hinders economic growth.[7] The growth-theoretic argument emphasises the so-called "resource-shallowing" effect: since labour force growth, by itself, cannot necessarily augment the availability of complementary productive resources, it will lead to declining capital/labour ratios and, therefore, to declining labour productivity because of the operation of the law of diminishing returns. The economic-demographic argument is that high population growth has a negative impact on the rates of saving and investment. A high rate of population growth increases the dependency ratio which, in turn, increases the share of consumption in national income (and thus reduces the share of saving in national income). A high rate of population growth also necessitates a diversion of investable resources from production-augmenting activities into building of social infrastructure (education and health facilities, housing, etc.).

These arguments, while theoretically plausible, have so far received little empirical support. Many theoretical counter-arguments have also been proposed. For example, a rapid growth of population, by causing a rapid expansion of the market, may induce technological innovations and may make it possible for economies of scale to be realised; it may worsen income distribution and therefore increase the rate of saving; resource-shallowing effect may be counterbalanced by organisational improvements; and so on. Perhaps a more relevant argument, consistent with the Lewisian dual economy paradigm, is that a rapid growth of population can neither affect the real wage rate in the modern sector nor alter the share of profit in national income; it therefore cannot affect the growth process. At any rate, a consensus is emerging now that no *a priori* relationship between demographic change and economic growth can be postulated.[8]

But if demographic change and economic growth are independent processes, then a rapid growth of population (and labour force) must slow down absorption of surplus labour and improvement in levels of living; this holds tautologically. It thus makes the Lewisian turning point a shifting target. But it does not account for stagnation or slow growth. And, besides, it does not explain growth of poverty and underemployment; slowing down improvement is not the same thing as causing deterioration. The real case for implementation of family planning programmes rests on the argument that it helps quicken the pace of improvement in living standards, given the pace of economic growth, and not on the expectation that it helps quicken the pace of economic growth itself.

[6] Some results of past research, which broadly support the views expressed here, are reported in Birdsall (1988).

[7] The details of the arguments and the references to the relevant literature can be found in Kelley (1988) and Birdsall (1988).

[8] See Horlacher and Mackellar (1988).

2.4 Conclusions

In the developing world, both population and labour force have been growing at historically unprecedented rates and the rates are still accelerating.

During the period under consideration, the labour force grew faster in more developed countries. In the next 15 years, however, the rate of growth of labour force will be roughly the same in low-, medium- and high-income countries. The challenge of employment creation will thus be growing in low-income countries and, to an extent, also in medium-income countries.

Demographic expansion and economic growth are relatively independent processes. This means that rapid demographic expansion slows down improvements in employment conditions and in levels of living. This is why implementation of population control measures is an urgent necessity in low-income countries.

3

Labour outmigration from agriculture

This chapter is concerned with three inter-related sets of issues. The first set of issues relates to the nature and pace of labour outmigration from agriculture in developing countries. The second set of issues relates to the factors which account for the general tendency of labour to move out of agriculture into non-agricultural activities in the course of economic growth. Finally, an attempt is made to explain the differences in the pace of labour outmigration across countries in terms of a specific set of hypotheses.

3.1 The pace of labour outmigration from agriculture: A general review

It is often thought that labour outmigration from agriculture in developing countries has been rather excessive. In popular perceptions, this is associated with the fact that urban slums have been steadily expanding in most countries. For many economists, this follows from the fact that in a large number of developing countries, state policies have discriminated against agriculture. It is arguable that had state policies been neutral, the pace of movement of labour out of agriculture would have been slower.[1]

Yet it is also widely held that underemployment in agriculture has not only persisted but has actually worsened in many countries.[2] The implication is that the pace of labour outmigration from agriculture has been too slow. Indeed, it is widely expected to remain so, since a common policy prescription today is that labour absorption in agriculture should be promoted through active policy interventions by the government.[3] This stands in sharp contrast with the view, dominant in the fifties and sixties, that the evolution of conditions of employment in agriculture depended on the growth of employment in the urban-industrial sector and the latter rather than the former constituted the proper domain of policy intervention.

[1] See Squire (1981).

[2] This is suggested by the available studies on rural poverty. See, for example, Khan and Lee, eds. (1984); Ghai and Radwan, eds. (1983); Jamal (1986) and Saith (1989a). The issue is explored in some detail in Chapter 4.

[3] See, for example, ILO/ARTEP (1980a, 1980b); Gooneratne (1982) and Booth and Sundrum (1985).

Table 3.1. Share of agriculture in labour force and its annual rate of change (median values)

	Share 1969 (%)	Annual rate of change (%)	
		1969-79	1979-85
SSA	84.0 (39)	−0.50 (39)	−0.74 (39)
AP	66.6 (15)	−0.83 (15)	−0.96 (15)
LAC	45.2 (21)	−1.79 (21)	−1.93 (21)
NENA	49.1 (11)	−2.70 (11)	−3.11 (11)
DC	69.2 (86)	−0.92 (86)	−1.02 (86)

Note. Figures in parentheses indicate the number of observations in each case.

Source. Appendix table III.

Evidently, it is not easy to avoid contradictions in judging the adequacy (or otherwise) of the observed pace of labour outmigration from agriculture in developing countries. There is no unique criterion and the use of different criteria leads to different judgements. The best we can hope to do is to develop a perspective which is useful for analytical and policy-making purposes. In what follows, we shall, first, place the recent experience of developing countries with regard to labour outmigration from agriculture in a wider historical perspective and, secondly, examine its consequences for the labour/land balance in agriculture. In the course of this analysis, we shall also have occasion to observe the effects of the recent economic recession on the migration process.

The particular index used to measure the pace of labour outmigration from agriculture is the annual percentage change in the proportion of labour force in agriculture. Two assumptions are needed to justify the use of this index. The first assumption is that the natural rate of growth of labour force is the same in all sectors so that agriculture's share in total labour force would have remained unchanged in the absence of outmigration. The second assumption is that when labour migrates out of agriculture, it necessarily goes into the non-agricultural sector within the country.

Neither of these assumptions can be expected to hold strictly in real-world situations. Fertility, mortality and labour force particpation rates can conceivably vary between agricultural and non-agricultural sectors. Also, when labour moves out of agriculture, it may go out of the labour force altogether (e.g. may migrate to another country). However, the importance of these factors is unlikely to be such as to render the index used here misleading in qualitative terms.

The country-level estimates are presented in Appendix table III while the regional aggregates are presented in table 3.1 above. The first point to be noted is the universality of the tendency of the share of agriculture in labour force to decline; the share declined in all the countries considered, albeit at different rates.

The second observation that can be made is that, roughly speaking, the pace of movement of labour out of agriculture was constant during the two periods considered. Given the lower initial values for the later period than for the earlier period, the observed higher rates of change in the later period actually imply a rough

Table 3.2. Initial share of agriculture in labour force and its annual rate of change in selected developed countries

Country	Initial share		Annual rate of change	
	Year	%	Period	%
Norway	1865	63.7	1865-1891	−0.92
			1891-1920	−1.04
			1920-1960	−1.58
Sweden	1860	64.0	1870-1910	−1.12
			1910-1960	−2.47
Finland	1880	71.2	1880-1960	−0.86
Spain	1877	70.5	1877-1964	−0.80
United States	1810	83.7	1810-1840	−0.92
			1840-1929	−1.22
			1929-1965	−3.58
Japan	1872	85.8	1872-1900	−0.67
			1900-1920	−1.31
			1920-1964	−1.54

Source. S. Kuznets, 1971, pp. 250-253.

constancy. This is an interesting fact since a large number of developing countries, mostly outside Asia-Pacific, faced grave economic difficulties in the later period. The economic recession apparently had very little impact on the process of labour transfer from agriculture into non-agriculture.

A third interesting feature emerges from a comparison of table 3.1 with table 3.2 which presents data on the pace of movement of labour out of agriculture observed in a few selected developed countries. At initial periods, it can be seen, the share of agriculture in the labour force in these countries was quite similar to that observed in many of today's developing countries, so that a comparison is legitimate. It is clear that in relation to the historical experience of today's developed countries, the pace of labour outmigration from agriculture in developing countries has been neither rapid nor slow. Indeed, in view of the fact that the rate of growth of labour force is much higher in developing countries than it was in today's industrialised countries at a comparable stage of development, the trends in developing countries appear quite remarkable.

But contemporary developing countries are late industrialisers and generally suffer from severe pressures of population and labour force on land.[4] The land/worker ratio is particularly unfavourable in the countries of sub-Saharan Africa and Asia-Pacific, as can be seen from the data in table 3.3. And it is precisely in these regions that the land/worker ratio has been declining rather rapidly; in sub-Saharan

[4] The contemporary industrialised countries, when they initiated modern economic growth, had a much higher land/worker ratio in agriculture than is observed in contemporary developing countries. See Kuznets (1965).

Table 3.3. Arable land per agricultural worker and its annual rate of change (median values)

	Arable land (ha), 1969	Annual rate of change (%)	
		1969-79	1979-85
SSA	1.18 (39)	−0.76 (39)	−1.21 (39)
AP	0.95 (15)	−0.82 (15)	−0.64 (15)
LAC	2.71 (21)	0.77 (21)	0.0 (21)
NENA	4.46 (11)	0.0 (11)	0.0 (11)
DC	1.49 (86)	−0.18 (86)	−0.83 (86)

Note. Figures in parenthesis indicate the number of observations in each case.
Source. Appendix table VII.

Africa, moreover, the rate of decline accelerated during 1979-85. In Latin America-Caribbean, land per worker increased during 1969-79 but remained unchanged during the subsequent period. In Near East-North Africa, land per worker remained unchanged during both periods. The significance of these facts becomes clear once we recognise that land per worker is an important indicator of underemployment in agriculture (as we shall see in Chapter 4) and that its variations over time result principally from variations in the number of workers in agriculture (because the land frontier cannot be easily expanded in most countries). Given the fact that agriculture, generally speaking, is a reservoir of surplus labour, the number of workers in agriculture is in the nature of a residual whose variations over time depend on the pace of labour outmigration from agriculture.

From this perspective, it thus appears that far from being excessive, the pace of labour outmigration from agriculture in the countries of sub-Saharan Africa and Asia-Pacific has in fact been too slow. Only in the case of Latin America-Caribbean and Near East-North Africa can the pace of labour outmigration from agriculture be regarded as adequate. Given this context, the fact that the experience of developing countries appears quite consistent with the historical experience of today's developed countries only serves to underline the immensity of the task facing the late industrialisers. They start with a larger stock of surplus labour in agriculture and have a higher rate of population growth and, consequently, need a faster rate of labour transfer to achieve a comparable pace of structural transformation. It may indeed be true that state policies in developing countries have discriminated against agriculture; it may also be important to search for policies which stimulate labour absorption in agriculture. But it is quite wrong to suggest, by way of justification, that the pace of labour outmigration from agriculture needs to be slowed down. In fact, an acceleration in the pace of labour outmigration is desirable and this means that stimulating employment growth in the non-agricultural sector remains a matter of central importance.

We have so far been concerned with the average picture which conceals considerable inter-country variations. In a subsequent section, we shall attempt to explain these variations in terms of certain hypotheses relating to the growth process.

Here it is of some interest to look more closely at the proposition that the economic difficulties in the post-1979 period had little impact on the pace of outmigration from agriculture. We have isolated, in table 3.4, a set of countries which were particularly affected by the post-1979 crisis. From the data on the rate of growth of non-agricultural GDP, the effects of the crisis are quite clear. It is also clear that in a majority of the countries the pace of labour outmigration from agriculture did not slow down in spite of the very sharp deceleration in the rate of growth of non-agricultural GDP. In a few countries, mostly in Latin America-Caribbean, there is an indication of a slight slowdown in the pace of labour outmigration, but the slowdown is quite insignificant in relation to the extent of deceleration in the growth of the non-agricultural sector.

The conclusion that the economic crisis had little impact on the pace of labour outmigration from agriculture, therefore, appears quite justified.[5] Workers were moving, in the post-1979 period, to a sector where average incomes were falling quite rapidly. Given that new entrants usually become marginal workers,[6] it hardly seems plausible to suppose that they were being induced by the lure of higher incomes. We shall consider the analytical implications at a later point in the chapter.

3.2 A search for causal explanations

The tendency of labour to move out of agriculture in the course of economic growth is, of course, one of the well-established generalisations in economics.[7] Three basic explanations have been advanced in the literature to account for this tendency. The first is the Engel's Law which states that as personal incomes rise, consumers spend a progressively lower proportion of expenditure on food and, correspondingly, a progressively higher proportion on non-food items. This means that the demand for food (an agricultural product) rises less rapidly than GDP while the demand for non-food items (largely manufactured products and services) rises more rapidly than GDP. The production structure adjusts to this changing structure of demand and a reallocation of labour occurs in the process.

The second explanation, which also emphasises changes in the structure of demand, points to changes in the composition of exports in the course of development.[8] At early stages of development, it is held, a country's exports consist mainly of agricultural products; but as growth occurs, the weight of manufactured products in exports gradually rises. This tendency, too, induces a reallocation of labour in favour of manufacturing.

[5] One issue that remains unresolved concerns the possible biases in the data. Population censuses are conducted every ten years and generation of annual data on population and labour force always involve projections. Some countries do conduct sample surveys in some of the inter-censual years; projections for these countries are fairly reliable. For other countries, however, the projections are based, of necessity, on the past trends and somewhat *ad hoc* future expectations. Judging the reliability of the data for 86 developing countries is a complex task which requires a separate study; ideally, a country-by-country analysis should be conducted. And it is not clear that, even after such an analysis has been undertaken, all doubts can be removed. For a proper confirmation of the findings reported here, we have to wait until the results of the next round of censuses are available. It is, however, worth noting that a recent study on sub-Saharan Africa also reaches a similar conclusion. See Jamal and Weeks (1988).

[6] See Yap (1977) and Williamson (1988).

[7] See Clark (1957), Kuznets (1971) and Chenery and Syrquin (1975).

[8] See Balassa (1979).

Table 3.4. Growth of non-agricultural GDP and share of agriculture in labour force (percentages)

Country	Non-agricultural GDP		Share of agriculture in labour force	
	1969-79	1979-85	1969-79	1979-85
Malawi	6.89	1.89	−0.80	−0.93
Lesotho	12.25	0.00	−0.42	−0.67
United Republic of Tanzania	5.03	0.90	−0.53	−0.53
Central African Republic	2.78	0.00	−1.34	−1.35
Togo	4.58	0.00	−0.50	−0.46
Sudan	7.31	1.66	−0.80	−1.46
Mauritania	2.41	0.00	−1.94	−0.81
Nigeria	7.94	−4.66	−0.40	−0.48
Mauritius	11.56	2.85	−1.34	−2.02
Côte d'Ivoire	7.58	0.00	−1.55	−1.51
Liberia	0.77	−3.13	−0.42	−0.54
Zambia	1.42	0.00	−0.47	−0.57
Philippines	6.37	0.00	−0.58	−0.96
Honduras	5.35	0.00	−0.70	−0.89
Ecuador	11.83	1.79	−2.63	−2.38
Guyana	3.20	−4.15	−1.76	−1.78
Brazil	8.79	1.61	−3.50	−2.58
Peru	4.26	0.00	−1.60	−1.42
Uruguay	3.16	−3.29	−1.64	−1.49
Trinidad-Tobago	5.10	−3.84	−5.64	−3.48
Argentina	2.26	−2.08	−2.06	−2.25

Source. Appendix tables III and V.

The third explanation points to the differential rates of growth of labour productivity in different sectors in the course of development. Historical experience shows that labour productivity in agriculture increases faster than that in industry and services. This serves to reinforce the effects of demand shifts.

These explanations, it may be noted, are meant to apply generally to all countries and not particularly to developing countries. Indeed, there are good reasons to doubt their relevance in the context of developing countries. When there is surplus labour in agriculture, labour outmigration does not imply a deceleration in the growth of agricultural or food output. Moreover, per capita food consumption in most developing countries is low and poverty is widespread. Under these conditions, if the demand for food increases less rapidly than incomes, then this is much more likely to be a reflection of worsening income distribution than of the Engel's Law (otherwise we would need to assume perverse behaviour on the part of the malnourished). In fact, a quite typical situation in developing countries is one where efforts to industrialise rapidly cause the demand for food to rise faster than supply leading to rising food prices and (consequently) worsening income distribution. The point is that

Table 3.5. Growth of food consumption (calories) and of GDP per capita (median values)

Region	Calories per capita per day, 1969	Annual rate of growth			
		Calories per capita		GDP per capita	
		1969-79	1979-85	1969-79	1979-85
SSA	2173 (39)	0.53 (39)	0.0 (39)	0.62 (30)	-1.94 (27)
AP	2223 (15)	1.31 (15)	0.52 (15)	3.16 (10)	2.77 (9)
LAC	2460 (21)	0.83 (21)	0.32 (21)	2.08 (17)	-2.01 (14)
NENA	2529 (11)	0.81 (11)	1.57 (11)	5.58 (7)	2.89 (6)
DC	2255 (86)	0.68 (86)	0.0 (86)	1.93 (64)	-0.99 (56)

Notes. Figures in parenthesis indicate the number of observations in each case. The abbreviations used throughout are as follows: sub-Saharan Africa (SSA), Asia-Pacific (AP), Latin America-Caribbean (LAC), Near East-North Africa (NENA) and Developing countries (DC).

Source. Appendix tables IV and VI.

it is difficult to see the observed changes in the production structure in developing countries as having been demand-driven; it is more plausible to suppose that changes in the production structure forced changes in the demand structure. It is one situation where, as people's food consumption level approaches saturation point, they shift their priorities to non-food items of consumption; it is quite another where the growth of food consumption is restrained because efforts to promote industrialisation bring about distributional changes. Economic growth in developing countries is associated with a rising share of profits in national income.[9] As a consequence, the share of consumption out of profits in total consumption also rises and consumption out of profits is directed primarily towards manufactured goods and modern services. Thus, so long as incomes are growing, the rate of growth of food demand can be expected to be lower than that of incomes in the aggregate, even if the consumption patterns of the agricultural population and of the urban workers remain unchanged. But the actual rate of growth of food consumption, in this perspective, is governed not by the rate of growth of incomes but by such factors as rate of growth of food production, international flows of food aid and government policies with respect to export and import of food.

The data presented in table 3.5 not only show the low levels of food consumption in developing countries but also suggest that it is best to assume the growth of food consumption to be independent of the growth of incomes. In the seventies, when growth was healthy and (presumably) the share of profits in national income was rising, the rate of growth of food consumption was indeed lower than that of per capita GDP in all regions. But in the eighties, food consumption remained stable in sub-Saharan Africa and grew at a respectable rate in Latin America-Caribbean, even though per capita GDP declined in both regions. In Near East-North Africa, moreover, the rate of growth of food consumption was much faster in the eighties than in the seventies, even though per capita GDP grew much faster

[9] Lewis (1954).

Table 3.6. Agricultural exports as percentage of total merchandise exports (median values)

Region	1969/71	1978/80	1983/85
SSA	65.4 (39)	57.9 (39)	46.6 (39)
AP	43.6 (15)	34.0 (15)	24.3 (15)
LAC	53.4 (21)	42.3 (21)	47.9 (21)
NENA	23.7 (10)	13.0 (10)	4.9 (10)
DC	59.4 (85)	38.3 (85)	35.1 (85)

Note. Figures in parentheses indicate the number of observations in each case.
Source. Appendix table IX.

in the seventies than in the eighties. The evidence, on the whole, does not suggest any systematic relationship between economic growth and the growth of food consumption.

As for the second explanation, there is some evidence to suggest that the share of agricultural exports in total exports has indeed been declining. The data in table 3.6, however, should be interpreted with caution. Since both total exports and agricultural exports were valued at current prices, the percentages reported in table 3.6 incorporate effects of changes in international relative prices. It is known that these moved in favour of agricultural commodities in the 1970s and against agricultural commodities in the 1980s. Thus the figures in table 3.6 probably understate the decline in the share of agricultural exports in total merchandise exports for the 1970s and overstate it for the 1980s. It is also to be noted that in the 1980s, the share of agricultural exports in total merchandise exports actually increased in the case of the Latin America-Caribbean region. The overall conclusion seems to be that the export structure changed in the predicted direction in the 1970s but not in the 1980s. The relevance of changing export structure in explaining labour outmigration from agriculture is thus not at all clear. At any rate, the question of causality remains; it may be more reasonable to explain the changes in export structure in terms of changes in production structure.

Comparisons of the growth of labour productivity in different sectors are not very meaningful in the context of developing countries. In particular, because of the preponderance of family enterprises and casual wage-labour in agriculture, it is very difficult to estimate labour productivity in agriculture that is conceptually similar to labour productivity in manufacturing where regular wage-employment predominates. At any rate, the available evidence suggests that the rate of growth of labour productivity has been slower in agriculture than in non-agriculture.[10]

The traditional explanations, therefore, do not appear to be particularly relevant in the context of developing countries. An alternative source of explanation is the standard dual economy model.[11] The prime mover in a dual economy is capital

[10] See Kuznets (1971), ch. IV.

[11] The literature is vast. Among the important studies are Lewis (1954, 1958, 1979); Fei and Ranis (1964); Gersovitz, Diaz-Alejandro and Rosenzweig, eds. (1982) and Kanbur and MacIntosh (1987).

accumulation in the modern sector – a sector which, in contrast to the traditional sector, uses capital in production and employs labour on the basis of a wage system. In a closed economy context, as capital accumulation occurs and the modern sector expands, labour moves out of the traditional agricultural sector (a major product of which is food) while leaving per capita food output unchanged. Thus, on the production side, while per capita industrial output rises steadily, per capita food output stays constant. On the demand side, capital accumulation is associated with a rising share of profits in GDP and a constant real wage rate. These factors ensure that per capita demand for food remains constant while the demand for industrial products rises steadily. These are, of course, schematic formulations and it is worth pointing out that neither a constancy of food output per capita nor a constancy of the real wage rate in the modern sector is strictly necessary for the argument. It is sufficient that (i) the level of growth of food output per capita is not affected by labour outmigration from the traditional sector and that (ii) the share of profits in GDP rises as the modern sector expands (i.e. the real wage rate rises less than labour productivity).

The reasoning, in this framework, flows from an assumed autonomy of the capital accumulation process which is based on productive use of hitherto underemployed labour drawn from agriculture. It denies autonomy to demand. It is not the change in the structure of demand which induces change in the production structure and hence leads to a reallocation of labour; rather it is the capital accumulation process which changes the production structure, induces labour outmigration from the traditional sector and changes the structure of demand.

Openness of the economy only increases the autonomy of the accumulation process since foreign savings may flow in, natural resources (e.g. minerals) can be converted into capital through trade and food can be imported. The basic results derived for the closed economy case, therefore, easily carry over into the open economy case. The model suggests that migration to the modern sector is regulated entirely by the demand of the modern sector for labour. However, there has been a tendency in the literature to identify non-agriculture as the urban-modern sector and this, we shall argue in the following section, has been a source of much confusion. The well-known Harris-Todaro model,[12] for example, starts from the premise that there is unemployment in the modern sector – a fact which the dual economy model had not taken into account. The model then proceeds to attribute the existence of unemployment in the modern sector to the fact that labour-incomes typically tend to be significantly higher in the modern sector than in the traditional sector. It can be shown that, under these conditions, the migration process will always generate a certain rate of unemployment in the modern sector.

It can be pointed out that this argument is based on a presumed optimising behaviour of individual labourers and, as such, is inconsistent with a standard assumption of the dual economy model, namely, that households constitute the basic unit of production and distribution in the traditional sector. If we suppose that it is the household income that is optimised, then the differential that counts is that between the average labour-income in the modern sector and the marginal contribution of a worker to household income in the traditional sector. So long as

[12] See Harris and Todaro (1970) and Todaro (1976).

there is surplus labour in the traditional sector, such a differential will exist whenever the average income per worker in the modern sector is positive (since the marginal contribution of a worker to household income in agriculture can be supposed to be zero).

But, fundamentally, if we suppose the non-agricultural sector to have a significant traditional segment, the basic premise of the argument disappears.[13] The unemployed (or the underemployed), it can be argued, are not in the modern sector but in the traditional non-agricultural sector. And it need not be supposed that labour migrates from agriculture to the modern sector. In the following section, we shall try to show that there are valid grounds for such arguments.

3.3 Employment in non-agriculture

The proposition that the non-agricultural sector in developing countries cannot be presumed to be equivalent to the modern sector of the dual economy model is not new. Much has been written about the urban informal sector and the rural non-farm sector – the two sectors which together constitute the traditional non-agricultural sector.[14] But perhaps because the urban informal sector and the rural non-farm sector have generally been treated separately and have been seen as appendages of the modern and the traditional sectors respectively, the analytical implications of their existence have gone largely unnoticed in the theoretical literature.

The obvious (and the only feasible) way of defining the modern sector is as that segment of the non-agricultural sector which employs labour exclusively on the basis of a wage system. Appendix table X presents the available data on the share of wage employment in total labour force in non-agriculture; in view of the fact that the number of cases for which data are consistently available is not large, it is not worth while presenting summary statistics for the regions. The data in Appendix table X highlights two facts. First, the non-agricultural sector itself is highly dualistic; the modern sector is only a segment of the non-agricultural sector and quite a small segment in many countries. Once we recognise this, it would seem natural to suppose that labour migrates from traditional agriculture to traditional non-agriculture and from the latter to the modern sector. In that case, it remains true that the modern sector draws labour from the traditional sector only in accordance with its needs and to talk about unemployment or underemployment in the modern sector is not very meaningful. Moreover, growth of employment in the modern sector influences labour outmigration from agriculture only indirectly, i.e. in so far as this leads to a reduction in the size of the labour force in traditional non-agriculture. For these reasons, differentials in labour incomes between the modern and the traditional sectors cannot have any influence on labour outmigration from agriculture. As we have already suggested, it is worth while for a worker to migrate from agriculture to traditional

[13] A similar argument is advanced by Jamal and Weeks (1988) in the context of sub-Saharan Africa.

[14] See, for example, Mazumdar (1975) and Saith (1989b).

non-agriculture so long as he can hope to earn a positive income in the latter sector because his marginal contribution to family income in agriculture is zero.[15]

The second fact, which also supports the conjectures above, is that while, in the seventies, the share of wage employment in non-agricultural labour force generally tended to rise, it generally tended to fall in the post-1979 period. This suggests not only that labour continued to migrate from agriculture to traditional non-agriculture even when modern sector employment was not growing (or was growing only very slowly) but also that labour thrown out of employment in the modern sector was taking refuge in the traditional non-agricultural sector.

The same facts also emerge from a comparison of Appendix table II with Appendix table X. In the seventies, wage employment grew at a faster rate than labour force in the non-agricultural sector in most countries. In the eighties, however, the rate of growth of the labour force was generally higher than that of wage employment; in fact, in quite a number of countries, wage employment recorded an absolute decline while the labour force in non-agriculture continued to grow. Thus the growth performance of the modern sector affected the relative size (in terms of employment) of the traditional non-agricultural sector but not necessarily the overall rate of growth of non-agricultural labour force. This, of course, is only another way of saying that there was no straightforward relationship between labour outmigration from agriculture and the growth performance of the modern sector.

Much attention has been paid to the growth of manufacturing employment in the literature. In particular, it has often been argued that the growth of manufacturing employment in developing countries has been too slow to reduce pressures on agriculture substantially. Two ideas are implicit in this argument: that the rate of growth of manufacturing employment has been slow and that the rate of growth of manufacturing employment has a major impact on labour outmigration from agriculture. We shall consider them in turn.

Two theoretical arguments are usually put forward to account for the alleged slow growth of manufacturing employment and both of these relate to issues of choice of technique. An old argument is that price distortions in developing countries – involving, typically, overpricing of labour and underpricing of capital – lead to a choice of inappropriately capital-intensive techniques and this keeps the actual growth of industrial employment below the potential.[16] A more recent argument is that developing countries, typically, have ignored comparative advantage (by pursuing protectionist policies) and this has restricted the growth of manufacturing employment since exportables tend to be more labour-intensive than non-exportables.[17]

[15] The available micro-level studies appear to indicate that average incomes of migrants in non-agriculture tend to be higher than what they earned in agriculture before migration. However, there are obvious methodological difficulties in establishing such a proposition and it is likely that the surveys (if any) carried out in the eighties will show rather different results. But even if the proposition is valid, it does not contradict the hypothesis stated in the text. It is also of some relevance to note that the few available studies on urban-rural remittances show that gross remittances are significant in Asia and sub-Saharan Africa but virtually non-existent in Latin America, while net remittances are insignificant everywhere. These facts, too, are consistent with the hypothesis stated in the text. See Yap (1977) and Williamson (1988).

[16] Cf. Morawetz (1974).

[17] See Krueger (1983, 1988) and Balassa (1984).

Table 3.7. Annual rate of growth of manufacturing employment (median values)

Region	1969-79	1979-85
SSA	3.09 (25)	0.0 (26)
AP	7.72 (11)	1.67 (11)
LAC	2.81 (17)	−1.23 (17)
NENA	4.35 (6)	6.62 (6)
DC	3.96 (59)	0.0 (60)

Note. Figures in parentheses indicate the number of observations in each case.
Source. Appendix table XII.

Table 3.8. Elasticities of wage employment (EWP) and total employment (ETP) w.r.t. output in manufacturing, 1969-79

SSA	*EWP*	*ETP*	*LAC*	*EWP*	*ETP*
Ethiopia	–	1.35	Bolivia	0.91	0.50
Lesotho	–	0.59	Ecuador	–	0.91
Burundi	1.03	0.83	El Salvador	–	0.50
Botswana	0.58	0.53	Nicaragua	0.91	0.99
Kenya	0.74	0.75	Brazil	–	1.00
Nigeria	–	0.75	Panama	–	0.93
Mauritius	2.39	1.73	Mexico	–	0.38
Côte d'Ivoire	–	1.02	Jamaica	–	0.62
Zimbabwe	1.09	0.85	Chile	–	0.30
Zambia	0.68	1.07	Trinidad-Tobago	–	1.53
Sierra Leone	–	0.40	Venezuela	–	1.63
Liberia	2.16	–	Colombia	–	0.91
Cameroon	2.54	–	Peru	1.24	–
			Paraguay	0.52	–
AP			*NENA*		
Indonesia	–	0.51	Jordan	–	0.53
India	0.57	0.78	Tunisia	–	0.90
Bangladesh	–	0.60	Libyan Arab Jamahiriya	0.79	–
Sri Lanka	0.95	2.31			
Philippines	–	1.47			
Korea, Rep. of	0.74	0.71			
Malaysia	–	0.93			
Pakistan	–	0.17			
Thailand	–	0.75			
Fiji	1.00	–			

Note. The ETPs were derived by fitting log-linear regression equations to the annual data on employment and GDP in manufacturing (all the values reported were statistically significant) while the EWPs were derived by dividing the rate of growth of employment by that of GDP.

These arguments ignore some typical problems faced by the late industrialisers. They must import technology from countries which are already highly industrialised and the range of choices with respect to labour coefficients is not wide. At any rate, the real issue concerns the rate of growth of employment and not employment per unit of capital at a point of time. Capital-intensive techniques may well facilitate a faster growth of manufacturing output and, therefore, of employment.[18] It has not been shown that capital intensity of techniques tended to rise over time. However, the choice of technology is, to a large extent, determined by the choice of products and here the second argument becomes relevant. Two points need to be noted in this context. First, the evidence on the allegedly higher labour-intensity of exportables (as compared to non-exportables) is not in fact very strong. There is no good reason, moreover, why we should confine attention to tradeables alone. Secondly, there are good reasons why developing countries adopt protectionist policies thereby emphasising the development of capital and intermediate goods industries. Historical experience shows that similar strategies were pursued with success by the late-industrialisers in Europe.[19]

The available evidence does not really support the view that the rate of growth of manufacturing employment has been slow. Table 3.7 shows that this was in fact quite rapid in the seventies both in relation to the historical experience of the industrialised countries and in relation to the rate of growth of labour force.[20] The rate of growth slowed down, and even became negative in many countries, in the eighties, but this, as we know, has been a period of general economic crisis. The very same tendencies were also observed in the case of growth of wage employment, as can be seen from Appendix table XII.

The estimates of employment elasticities in manufacturing, presented in table 3.8 do not suggest that employment growth has been, as a rule, restricted by inappropriate choices of technique. In most countries, output growth was achieved through a combination of employment growth and productivity growth, but the former was the more important factor (this is indicated by the fact the elasticity values generally lie between 0.5 and 1.0). In a few countries, employment growth was in fact excessive (elasticity values exceeding 1.0) in the sense that it was accompanied by declining labour productivity. Only in very few cases was the elasticity low (less than 0.5). The overall conclusion seems to be that technological choices did not have a major effect on employment growth in manufacturing. We should add that, for this analysis, we have deliberately chosen to consider only the period 1969-79 since, in the 1980s, the economic crisis rendered the relationship between employment and output erratic and unstable.

Another point, which is perhaps worth noting, is that the data in table 3.8 do not in fact suggest that the employment elasticity was systematically higher in countries pursuing export-oriented industrialisation than in those pursuing import-substitutive industrialisation. A comparison of India (a good example of

[18] Cf. Dobb (1963) and Sen (1968).

[19] A revealing analysis is available in Gerschenkron (1962).

[20] Similar conclusions are drawn in Squire (1981), which also presents evidence on the historical experience of the industrialised countries.

Table 3.9. Regression results, dependent variable: Rate of growth of wage employment in non-agriculture

Type of regression	Period	Constant	RGWPM	R^2	n
Linear	1970-79	3.49 (4.45)	0.36* (4.01)	0.47	20
Linear	1979-85	1.03 (1.34)	0.29** (2.52)	0.20	27

Notes. RGWPM = rate of growth of wage employment in manufacturing. * Significant at 1 per cent. ** Significant at 5 per cent. Figures in parentheses are t-statistics.

import-substitutive industrialisation) and the Republic of Korea (a model example of export-oriented industrialisation) is of some interest in this context. As can be seen, the "total employment elasticity" was roughly the same for the two countries while the "wage-employment elasticity" was higher in the Republic of Korea. These are tentative findings which need to be confirmed through further research. But they suggest that import-substitutive industralisation may appear less labour-intensive than export-oriented industrialisation if only the modern sector is considered, but not so if the total employment effect is taken into account.

It has to be noted at this point that the share of manufacturing in wage employment in the modern sector is generally quite small; it is around 20 per cent in most cases as Appendix table XI shows. Much of the wage employment is in fact in services. In these respects, the developing countries' experience differs from that of the industrialised countries in their early stages of development. The share of manufacturing in non-agricultural employment was significantly higher (and the share of services correspondingly lower) in the industrialised countries than it is in today's developing countries.[21] It is this type of evidence which inspires arguments relating to techniques being prematurely capital-intensive in developing countries.

The fact that the manufacturing sector accounts for a small proportion of the modern sector employment does not imply, however, that it is only of minor importance as far as employment generation is concerned. For as the results of regression exercises in table 3.9 show, growth of employment in the entire modern sector is strongly and positively related to that in manufacturing. And since the growth of wage employment in manufacturing is primarily a function of manufacturing output, it can be inferred that the growth of the manufacturing sector plays a central role in generating employment in the modern sector. In other words, the manufacturing sector can be regarded as a leading sector.

It still does not follow, however, that there is a straightforward relation between labour outmigration from agriculture and growth of manufacturing output. As argued earlier, employment growth in the modern sector influences labour outmigration from agriculture only through its effects on the relative size of the labour force in traditional non-agriculture. This means that the rate of growth of

[21] See Kuznets (1971) and Squire (1981).

manufacturing output has to be above a certain threshold for it to have an impact on labour outmigration from agriculture. This is the hypothesis which we attempt to test with cross-country data in the following section.

3.4 A cross-country analysis

We seek to explain empirically, in this section, the variation of the percentage rate of decline in the share of agriculture in labour force across countries. The two explanatory variables chosen for the purpose are: the percentage rate of growth of manufacturing GDP and the overall rate of growth of the labour force. The first of these is intended to represent the growth of the modern sector; the rationale behind this choice has already been explained.

The second explanatory variable – the overall rate of growth of the labour force – is introduced as an imperfect proxy for the increase in the pressure of population and labour force on land in the agricultural sector. The assumptions involved are as follows:

(i) that in the absence of outmigration, the rate of growth of the labour force in agriculture would have been the same as that in non-agriculture;

(ii) that the land-frontier in most countries has reached the point of near-exhaustion (this assumption is verified in Chapter 4); and

(iii) that changes in social and technical relations of production in agriculture have not been so drastic as to cause a sudden exodus of labour from agriculture.[22]

It needs to be recognised that the percentage rate of decline in the share of agriculture in labour force is quite sensitive to the value of the initial share. It is also plausible to suppose that the higher the initial share, the slower is the pace of movement out of agriculture. This is because a high share of agriculture in labour force signifies a low level of development and, therefore, a low capacity for capital accumulation in the modern sector. In view of these considerations, the initial share of agriculture in labour force also needs to be introduced as an explanatory variable.

Thus we are led to postulate a cross-country empirical relationship of the type:

DS = f (IS, YM, L)

where DS – rate of decline of agriculture's share in labour force;
IS – initial share of agriculture in labour force;
YM – rate of growth of manufacturing GDP;
L – rate of growth of the labour force.

[22] There are, of course, a few countries where development of capitalist farming and associated mechanisation may have had the effect of forcing labour out of agriculture. There are also a few countries where redistributive land reforms may have produced the opposite effect. The assumption is that such cases are not numerous enough to introduce systematic biases.

Table 3.10. Regression results, dependent variable – DS

Type	Sample	Constant	IS	YM	L	R^2	n
Linear	69-79	4.22	−0.05* (−6.65)	0.10* (3.42)	−0.08 (−0.47)	0.50	61
Linear	79-85	3.16	−0.04* (−10.60)	0.04* (2.89)	0.24** (1.90)	0.73	61
Linear	69-79a	4.35	−0.05* (−6.79)	0.13* (3.50)	−0.08 (−0.46)	0.54	53
Linear	79-85a	3.12	−0.05* (−8.34)	0.05** (1.77)	0.38** (2.00)	0.81	36
Log-linear	69-79a	6.01	−1.72* (−7.70)	0.55* (3.89)	0.14 (0.64)	0.59	53
Log-linear	79-85a	4.14	−1.28* (−5.68)	0.14 (0.94)	0.85** (1.87)	0.69	36

Notes. 1969-79 and 1979-85 are the two periods for which countrywise growth rates were calculated and used in the regression. 1969-79 and 1979-85 refer to samples of countries for which all the variables could be estimated. 1969-79a and 1979-85a refer to samples of countries which showed positive values for all the variables. IS pertains to 1969 for the period 1969-79 and to 1979 for the period 1979-85.
Figures in parentheses are t-statistics. * Significant at 1 per cent. ** Significant at 10 per cent.
n = number of observations.

The expected signs of the derivatives follow from the discussions in the preceding sections. We expect:

(i) that the higher the initial share of agriculture in labour force, the lower is the rate of outmigration of labour from agriculture ($f'_{IS}<0$);

(ii) that the higher the rate of growth of manufacturing output, the higher is the rate of outmigration from agriculture ($f'_{rm}>0$); and

(iii) that the higher the rate of growth of the labour force, the lower is the relative absorption capacity of agriculture and, therefore, the higher is the rate of outmigration from agriculture ($f'_L>0$).

The results of regression exercises, based on the specification above, are reported in table 3.10. In statistical terms, these results are quite acceptable. The most interesting point, suggested by these exercises, can be derived from a comparison of the results for the two periods. The rate of growth of the labour force turns out to be a significant explanatory variable only for the period 1979-85 when the performance of the manufacturing sector was poor in most developing countries. The rate of growth of manufacturing output operates as a major determinant of the pace of outmigration from agriculture only when the former is high. When the manufacturing sector stagnates, this does not stop outmigration from agriculture because agriculture is simply unable to absorb the incremental labour force.

These results confirm what has already been argued. Given the agrarian conditions in developing countries, the rate of growth of the labour force already determines a certain base rate of movement of labour out of agriculture. This rate is quite independent of both the rate of growth of manufacturing output and the intersectoral differentials in average incomes. Furthermore, this rate implies a base rate of growth of non-agricultural labour force and, therefore, a base rate of growth of manufacturing output that is needed if the distribution of the non-agricultural labour force between the modern and the traditional non-agricultural sectors is to remain unchanged. A rate of growth of manufacturing output that is below the base rate implies an expansion of the traditional non-agricultural sector. It is only when the actual rate of growth of manufacturing output is above the base rate that the former becomes a determinant of the pace of outmigration from agriculture. Agrarian conditions in this scheme serve as a shift parameter for the base rate of movement of labour out of agriculture. Agrarian reforms or technological development change the base rate for any given rate of growth of the labour force.

On this interpretation, it can be said that while in the seventies the rate of growth of manufacturing output was generally above the base rate, it fell below the base rate in most countries in the eighties. The fact that the pace of outmigration from agriculture was roughly similar in the two periods suggests that the base rate rose. In some countries, this could have been due to technological change in agriculture, but the more general reason was the acceleration in the rate of growth of the labour force.

3.5 Conclusions

Viewed in an historical perspective, the pace of labour outmigration from agriculture in developing countries appears to have been neither too slow nor too rapid. But when the specificities of the conditions in developing countries are taken into account, it has to be concluded that the pace has, in fact, been too slow in sub-Saharan Africa and Asia-Pacific. In this context, promotion of labour absorption in agriculture and creation of non-agricultural employment have to be simultaneous and should not be viewed as alternatives.

The evidence shows that the crisis of industrialisation, encountered in many developing countries in the eighties, had little impact on the pace of labour outmigration from agriculture. This suggests that intersectoral shifts of the labour force do not necessarily indicate genuine structural change.

It also casts doubt on the validity of many of the received theories concerning the forces governing intersectoral movements of labour, namely Engel's Law, changes in the structure of exports, differential growth of labour productivity in different sectors and the existence of differentials in labour incomes among sectors. An examination of the relevant evidence also shows them to be of little relevance.

The identification, implicit or explicit, of non-agriculture as the modern sector in the development literature has been a source of a great deal of confusion. It is quite misleading, for example, to talk about unemployment (or underemployment) in the modern sector or to suppose that labour migrates from agriculture to the modern sector.

The focus on the issues of choice of technique in manufacturing has also created some misleading impressions. Late industrialisers indeed adopt less labour-intensive techniques than early industrialisers. But this only means that for a given share of manufacturing in GDP, the share of manufacturing in the labour force is lower in late industrialisers than in early industrialisers. It does not have any implication for the relation between the growth of manufacturing employment and that of manufacturing output. The evidence shows that the employment elasticity of output is fairly high in the manufacturing sector of developing countries. The problem has been the sustenance of a rapid growth of the manufacturing output.

Besides, the contribution of the manufacturing sector to employment growth should not be judged in terms of the employment growth in the manufacturing sector alone. The growth of wage-employment in non-agriculture as a whole (i.e. the growth of modern sector employment) depends on the growth of the manufacturing output.

The overall perspective which emerges from the analysis in this chapter is as follows. Because of the near-exhaustion of the land-frontier, agriculture cannot absorb much of the incremental labour force even when land-augmenting technological change is significant. A certain rate of outmigration of labour thus occurs, irrespective of the growth performance of the non-agricultural sector, so long as labour can expect to earn a positive income in the non-agricultural sector. At a conceptual level, it can be supposed that labour moves from agriculture to traditional non-agriculture and from the latter to the modern sector. The growth of the modern sector influences the pace of outmigration from agriculture only when the former is high; otherwise, it merely determines the relative size of the labour force in traditional non-agriculture.

4

Underemployment in agriculture

This chapter is concerned with two issues. First, it seeks to assess the trends in agricultural underemployment in developing countries during 1969-85. This is not an easy task in view of the paucity of reliable statistical information on employment or underemployment in agriculture. However, collection of such information is inherently difficult and resource-intensive and, as such, the information base is unlikely to improve very much in the foreseeable future. It is necessary, therefore, to look for proxy variables. It is argued here that reasonable proxies can be found and that the broad tendencies relating to underemployment can be identified. Secondly, the chapter attempts to relate the trends in underemployment in agriculture to the patterns of agricultural growth and of outmigration of labour from agriculture. The evolution of agricultural underemployment is thus viewed not merely as a sectoral problem but in the context of the overall process of economic development.

4.1 Conceptual problems and difficulties of measurement

The existence of surplus labour or underemployment in agriculture has always been taken as a stylised fact in development economics. In fact, this was originally regarded as a particular advantage enjoyed by the developing countries. In these economies, it was thought, industry could potentially draw labour out of agriculture at a constant wage without affecting agricultural output. Surplus labour not only ensured that there was no labour constraint on industrialisation, it also represented a saving potential.[1] Such optimism, of course, no longer prevails and agricultural underemployment is now viewed as a major problem. Nevertheless, the concept retains its centrality in analytical discussions of development issues.

Yet the notion of underemployment has often appeared shrouded in conceptual ambiguities to economists. Some have attempted to deny altogether the existence of underemployment;[2] others have devoted considerable efforts to debating a variety of issues relating to it.[3] This is no place to go into these debates, but a few points emerging from them may be usefully noted.

[1] The ideas originate from the works of Lewis (1954) and Nurkse (1953) and have spawned a vast literature.

[2] See, for example, Schultz (1964).

[3] A brief review of the debates is available in Sen (1975).

Table 4.1. Incidence of wage labour in agriculture, c. 1980

Region	Wage labourers as percentage of agricultural workers (median value)
SSA	2.2 (7)
AP	18.2 (10)
LAC	30.8 (16)
NENA	23.0 (4)
DC	20.2 (37)

Note. Figures in parentheses indicate the number of observations.
Source. Appendix table XIII.

The idea that workers can be withdrawn from agriculture without affecting agricultural output does not imply that the marginal productivity of employed labour is zero in agriculture. Rather, it means that withdrawal of workers does not affect the total labour input employed in agriculture. Underemployment, therefore, refers to the fact that agricultural workers work fewer hours per day or fewer days per year than they desire at the given level of returns to labour.[4]

This situation arises because a principal factor of production, namely land, is in short supply and because there are only limited possibilities of substitution between land and labour.[5] Neither of these features, however, should be interpreted as purely technical in nature. The same level of land per worker may mean different levels of land-scarcity in two countries with different institutional characteristics (e.g. tenurial conditions). Similarly, substitution possibilities are determined partly by the technology in use and partly by the institutional framework of agriculture.[6]

These arguments usually presuppose that peasant farming predominates in agriculture. In many developing countries, however, the use of wage labour in agricultural production is quite extensive. It can be seen from table 4.1 that wage-labourers constitute a significant proportion of agricultural workers in all regions except sub-Saharan Africa. But this fact does not render the arguments above any less relevant. For the bulk of the wage labour in agriculture in most countries is in fact employed on a casual, daily basis. This is a well-known fact which is also brought out by the data presented in Appendix table XIII; these show that only a small proportion of the wage-labourers in agriculture are in regular wage-employment. Because of this, a situation arises where wage labourers, like peasants, are neither fully employed nor fully unemployed. Wage labourers, moreover, are not necessarily landless labourers but often are members of peasant households owning or operating small farms.

[4] For a rigorous formulation of these arguments, see Sen (1966).
[5] See Eckaus (1955).
[6] The works of Boserup (1965, 1981) are of interest in this context.

The fact that underemployment among wage labourers coexists with a positive wage rate in agriculture has led some economists to argue that it is the downward inflexibility of the wage rate which causes underemployment. In accounting for underemployment, according to this view, the inflexibility of the wage rate must be explained. The most important and interesting explanation, advanced in the literature, is known as the "efficiency wage hypothesis".[7] The argument is that there exists a functional relationship between the work-efficiency of a worker and the wage which determines the nutritive value of his food intake. Starting with a very low level of wage, work-efficiency may rise as the wage rises up to a point at which, in effect, labour in efficiency units becomes cheaper as the wage rises. It is then in the employers' interest to pay a wage that maximises the work-efficiency of the individual workers. This then explains why the wage rate may be inflexible downwards.

The problem with this explanation is that it is relevant only for regular, year-round wage employment. In the case of casual labourers, no single employer can possibly ensure a desirable level of nutritional intake for a labourer whom he employs only a few days in a year.[8] It might in fact be more fruitful to formulate the problem differently. Instead of supposing that the inflexibility of the wage rate causes underemployment among wage labourers, we may ask why a casual labour system exists at all. In our view, the system is a manifestation of two underlying features – existence of surplus labour and incomplete separation of labourers from means of production – both of which have to be understood in terms of historical processes of differentiation of the peasantry. The system cannot function unless there is chronic excess supply of labour, whatever the wage. The right question to ask is why the wage rate is what it is. There may well be no unique answer appropriate for all situations. For analytical purposes, what is important to realise is that a change in the demand for labour changes the degree of underemployment but not the wage rate.

The brief discussion above, while clarifying the conceptual basis of the notion of underemployment, also underlines the difficulties of monitoring it. The variables usually measured, such as the rate of unemployment, cannot be used as indicators of either the extent or the trends in underemployment. An adequate measurement would require carefully designed year-round surveys which identify workers willing to work for longer periods at the existing level of labour returns and the details of work presently done by each of the workers. Such surveys are much too costly to be regularly undertaken by the developing countries.

In the literature, the most widely used method of estimating underemployment has focused on the stock of unutilised labour. The standard procedure is to set a norm for full-time employment, to use this along with labour force data to derive an estimate of labour supply and then to subtract from this an estimate of current demand, which is simply the amount of labour deemed to be required to produce the given products in agriculture (usually based on small-scale surveys of

[7] See Bliss and Stern (1978).

[8] For a discussion in the context of India, see Ghose (1979).

Table 4.2. Rural poverty and land distribution

Region	% of rural population below poverty line, c. 1977 (median)	Gini coefficient of land distribution, c. 1975 (median)
SSA	56 (21)	0.43 (17)
AP	40 (13)	0.57 (10)
LAC	43 (11)	0.84 (13)
NENA	25 (5)	0.69 (5)
DC	50 (50)	0.62 (45)

Note. Figures in parentheses indicate the number of countries for which data were available.

Source. Appendix table XV.

cropwise labour requirements).[9] Clearly, this gives a measure of idle manpower rather than of underemployment; for, even the workers (adult members of landlord households, for example) who are not willing to work at the existing level of labour returns are counted as underemployed.

An alternative – our preferred alternative – is to focus on poverty.[10] In situations where property incomes are important, lack of employment does not necessarily mean poverty. But precisely for this reason, the category of poor is likely to exclude those underemployed who are unwilling to work at the existing level of labour returns. On the other hand, the category of poor may include people who are working full-time but earn extremely low returns for their labour. But the proportion of such people, generally speaking, is unlikely to be very significant. At any rate, to regard a person who works full-time but is poor as fully employed is to assign unacceptable connotations to the term "employment".

Having said this, it is perhaps appropriate to emphasise the need for empirical research designed to establish the nature of the relationship between underemployment and poverty. While *a priori* reasoning suggests a universal positive relationship, its strength and significance can be expected to vary across countries and it is of some importance to identify the factors which cause such variations. In principle, this type of research can be done on the basis of small-scale surveys and would fill an important gap in our knowledge.

9 For brief reviews of such studies, see Sen (1975), Turnham (1971) and Squire (1981).

10 Such an approach has been advocated by the ILO. See ILO (1970, 1972, 1974). On the other hand, Sen argues: "To identify unemployment with poverty seems to impoverish both notions since they relate to two somewhat different categories of thought." (Sen (1975), p. 38). At a general level, the argument is obviously valid. But, as pointed out in the text, there is a good case for identifying unemployment with poverty in the context of the agricultural sector.

4.2 Underemployment in agriculture: Magnitudes and trends

For the purpose of this study, we accept the incidence of poverty as a reasonably good proxy for underemployment and can now look at some data. Table 4.2 presents some available estimates of the incidence of rural poverty in a summary form. The general limitations of such poverty estimates are well-known and need not be repeated here.[11] A particular limitation of the estimates used here arises from the fact that the income level of the forty-sixth percentile of the rural population in India has been used as the uniform poverty line for all countries; there can obviously be perfectly reasonable objections to this methodology. Another point that must be noted is that the estimates refer to the rural population rather than to the agricultural labour force. As such, the estimates can only be treated as very rough indicators of underemployment in agriculture.

Even so, the estimates leave little scope for doubting that the problem of agricultural underemployment is very serious; around 50 per cent of the rural population were found to be in poverty even in the late seventies and a roughly similar percentage of the agricultural labour force could be assumed to be underemployed. It comes as no surprise that the level of underemployment appears to be highest in sub-Saharan Africa. What is perhaps more surprising is that the level of underemployment is of the same order in Latin America-Caribbean as in Asia-Pacific, even though arable land per agricultural worker is almost four times as high in the former region as in the latter. This provides a good illustration of how land-scarcity and poverty (and underemployment) are in part reflections of the institutional features of agrarian economies; land distribution in Latin America-Caribbean is much more unequal than that in Asia-Pacific. Indeed, the data in table 4.2 clearly suggest that while underemployment in Asia-Pacific and sub-Saharan Africa results, to a large extent, from land-scarcity, that in Latin America-Caribbean and Near East-North Africa is due primarily to inequalities in land distribution.

While table 4.2 gives some idea of the seriousness of the under-employment problem in developing countries, poverty indices cannot be used to study the trends in underemployment simply because time-series data on poverty are available for only a very few countries. Thus we are again confronted with the problem of finding proxies.

Some country studies have shown that output per worker in agriculture is a good indicator of employment per worker and as such relates strongly to poverty.[12] There is no reason why such a relationship should not obtain at a cross-country level. However, in any particular country, the institutional features of the agrarian economy can be taken as given and do not need to be explicity introduced in studying the relationship between output per worker and poverty. But at a cross-country level, the differences in institutional features across countries obviously cannot be left out of account.

[11] For a theoretical demonstration, see Kakwarni (1989). A good discussion of the mismatch between different sets of poverty estimates and between poverty estimates and other indicators of level of living, is available in Saith (1989a).

[12] See, for example, Ghose (1989).

Table 4.3. Regression results, dependent variable: ln (RP)

	Constant	ln (LPR)	ln (G_1)	ln (G_2)	ln (G)	R^2	n
1.	7.93	−0.63*	0.98**	–	–	0.59	17
	(8.18)	(−4.37)	(2.72)				
2.	7.09	−0.53*	–	0.70**	–	0.54	17
	(8.53)	(−3.99)		(2.31)			
3.	6.88	−0.49*	–	–	0.61**	0.48	27
	(10.15)	(−4.67)			(2.49)		

Notes. RP = incidence of rural poverty, c. 1977. LPR = agricultural GDP per agricultural worker in 1970 US$, average for 1976-78. G_1 = given coefficient of land distribution, c. 1970. G_2 = Gini coefficient of land distribution, c. 1980. G = G_1 and G_2 pooled together. * Significant at 1 per cent. ** Significant at 5 per cent. Figures in parentheses are t-statistics.

With these considerations in mind, we have carried out some regression exercises the results of which are presented in table 4.3. The Gini coefficient of land distribution has been used in these exercises as a composite indicator (admittedly imperfect) of the institutional features of agriculture. The results do support the idea that even at a cross-country level, output per worker is a reasonably good indicator of employment per worker; a higher level of output per worker indicates, ceteris paribus, a lower level of poverty and, therefore, a lower level of underemployment. On the other hand, a higher inequality of land distribution is associated either with a lower level of employment per worker or with a lower level of returns to labour or with both; hence the positive relationship between the poverty indices and the Gini coefficients.

These results provide some basis for supposing that output per worker is an appropriate variable for studying trends in underemployment. Since, in general, land distribution changes very slowly over time, the direction of change in output per

Table 4.4. Annual percentage growth of agricultural GDP per agricultural worker (median value)

Region	1969-79	1979-85
SSA	0.39 (32)	−1.30 (32)
AP	1.47 (14)	2.03 (14)
LAC	2.63 (20)	0.94 (20)
NENA	4.06 (9)	0.74 (9)
DC	1.97 (75)	0.00 (75)

Note. Figures in parentheses indicate the number of observations.
Source. Appendix tables II and IV.

Table 4.5. Trends in output per worker and real wage rates in agriculture, selected countries

Country	Annual rate of growth (%)			
	Output per worker		Real wage rate	
	1969-79	1979-85	1969-79	1979-85
Malawi	2.92	-1.64	-3.39	-0.71
Kenya	1.66	-0.11	1.38	-5.46
Ghana	-1.76	-1.51	-	-6.91
Mauritius	-0.31	-1.28	4.21	-
Zimbabwe	-2.21	-2.09	3.04	8.21
Bangladesh	0.75	0.63	-2.00	-0.62
Myanmar	2.60	4.91	-11.38	-3.32
India	0.46	2.08	0.00	2.99
Pakistan	0.08	0.75	3.93	16.49
Indonesia	3.16	2.57	0.00	-
Malaysia	3.99	2.48	1.74	-
Sri Lanka	0.79	2.51	6.52	-3.39
Philippines	2.88	0.66	-	-7.69
Korea, Rep. of	4.29	4.93	8.63	2.61
Fiji	1.21	-0.61	5.19	-8.49
Colombia	3.43	1.02	-	-1.16
Guatemala	-	-	-3.33	-
Guyana	-0.33	-1.32	-3.14	-
Nicaragua	2.85	-1.84	-4.41	-
Brazil	4.00	3.63	5.19	-
Costa Rica	2.02	1.54	5.49	-4.22
Panama	2.41	1.62	-0.64	2.04
Mexico	0.71	1.56	0.08	-4.07
Uruguay	1.50	0.85	-3.53	-6.48
Chile	3.86	2.41	-	-1.19
Argentina	3.94	3.18	-6.01	-
Egypt	1.97	0.74	-	-7.66

Source. Appendix tables II, IV and XIV.

worker should give a good idea of the direction of change in underemployment in any given country or region.[13]

The data in table 4.4 suggest the following conclusions. During 1969-79, agricultural underemployment declined in all regions, although the decline was not very significant in the countries of sub-Saharan Africa. Not surprisingly, the fastest decline occurred in the countries of Near East-North Africa whose oil revenues recorded, explosive growth during the seventies. During 1979-85, however, only in the

[13] Labour productivity in agriculture as an index of levels of living of the agricultural population is commonly used. See Bairoch (1975), Hayami and Ruttan (1985) and Timmer (1988).

Table 4.6. Some indirect indicators of changes in agricultural underemployment (median values)

Indicator		SSA	AP	LAC	NENA	DC
Arable land per agricultural worker, in hectares	1969	1.18	0.95	2.71	4.46	1.49
	1979	1.10	0.86	2.77	5.17	1.44
	1985	1.07	0.81	2.86	5.23	1.32
Percentage of arable land irrigated	1969	0.6	16.9	4.9	9.5	4.4
	1979	1.4	21.2	8.4	10.6	6.3
	1985	1.4	25.1	8.6	11.6	6.5
Fertilisers (kg) used per hectare	1969	1.6	14.3	26.8	5.6	5.0
	1979	2.8	44.4	50.3	22.7	15.0
	1985	5.8	59.2	39.1	38.4	18.6
Tractors per thousand hectares	1969	0.4	0.6	3.8	1.8	1.0
	1979	0.5	2.1	4.9	4.5	1.8
	1985	0.6	3.1	5.7	6.8	2.4

Note. The data on arable land per agricultural worker can give a misleading impression about the time-trends. Reference should be made to table 3.3, Chapter 3.

Source. Appendix tables VII and VIII.

countries of Asia-Pacific did underemployment continue to decline at a significant rate.[14] In the countries of sub-Saharan Africa, agricultural underemployment increased quite sharply. In the case of Latin America-Caribbean and Near East-North Africa we can conclude that the level of agricultural underemployment did not change significantly.

Table 4.5 presents data on the trends in output per worker and in real wage rates in agriculture for selected countries. Two points can be made on the basis of these data. First, it is clear that no particular relation between output per worker and real wage rate can be postulated. This is consistent with the idea that the real wage rate in agriculture is affected by the movements in relative prices rather than by the conditions of labour demand.[15] Secondly, when the two types of trends are considered together, they suggest some modifications of the conclusions (drawn above) about the trends in underemployment. In the countries of sub-Saharan Africa, the output per worker and the real wage rate tended to move in oppostie directions during 1969-79 and both tended to decline during 1979-85; thus the conclusion that underemployment did not change much during the first period and definitely increased during the second period remains valid. In the case of Asia-Pacific, the real wage rate declined in quite

[14] The trend growth of labour productivity in Asian agriculture appears to have been sustained over a long period. See, in this context, Hayami and Ruttan (1985).

[15] cf. Ghose (1989).

a few countries during 1979-85, and so the growth of output per worker overstates the decline in underemployment for this period. In the case of Latin America-Caribbean, the data suggest that the decline in underemployment could not have been very significant during 1969-79 (the real wage rate tended to decline in quite a few countries) and, during 1979-85, underemployment probably increased.

Finally, we can examine some data relating to a few indirect indicators presented in table 4.6. We know that employment per worker over time in any given situation would vary in the same direction as the amount of arable land per worker, the proportion of arable land irrigated and the amount of fertilisers used per unit of arable land.[16] Tractorisation can be labour-augmenting and labour-displacing at the same time and its net effect on employment is ambiguous.

Agriculture in sub-Saharan Africa remains basically traditional; arable land per agricultural worker has been declining and technological progress has been too slow to counteract the negative impact on employment per worker. In Asia-Pacific, technological progress has been quite impressive and of a labour-augmenting variety. In spite of the decline in land per worker, therefore, employment per worker is most likely to have increased. In Near East-North Africa, technological progress was rapid during 1969-79, while land per worker remained roughly stable. These developments do not seem to have slackened during 1979-85 and our earlier conclusion (that underemployment did not decline during 1979-85) requires modification. In Latin America-Caribbean, land per worker increased during 1969-79 but remained roughly stable during 1979-85. Technological progress was quite rapid in the seventies, but suffered a setback in the post-1979 period. Our earlier conclusions about the trends in agricultural underemployment thus remain valid.

Before concluding this section, it is useful to emphasise a few particular points. It is well-known that the countries of Asia-Pacific have not been much affected by the post-1979 global recession and debt crisis. This is evident even in such areas as agricultural underemployment and poverty; long-term trends have been maintained in the post-1979 period. In sub-Saharan Africa and possibly also in Latin America-Caribbean, on the other hand, the problems of agricultural underemployment and rural poverty worsened significantly in the post-1979 period and this could not be unrelated to the global recession and debt crisis. It has to be said, however, that the problems of agricultural underemployment and rural poverty did not show significant improvements in sub-Saharan Africa even in the pre-1979 period – a period of boom in the world economy. The problem obviously has a long-term character in sub-Saharan Africa.[17]

[16] Strictly speaking, these variables increase labour demand. But, as noted earlier, in a situation where underemployment is widespread, employment per worker is demand-determined.

[17] See also Timmer (1988) on this point.

4.3 Trends in agricultural underemployment: Some further explorations

The trends in underemployment, identified in the preceding section, obviously reflect the combined effect of the growth of the agricultural labour force and that of agricultural output. To take the analysis a step further, it is useful to ask which of the two variables was of primary importance in generating the observed trends in underemployment.

Of course, to ask this question is to assume that the two variables are not related. This is a correct assumption, since labour is not a constraining factor in agricultural production, at least not in the majority of the countries under study. The rate of growth of the agricultural labour force depends on that of the total labour force (i.e. on demographic factors) and the rate of outmigration from agriculture. Neither of these has much to do with agricultural growth which results from extension of cultivated areas and land-augmenting technological change. These changes, in turn, result from investment, which depends partly on state policy and partly on institutional features of agriculture.

The relevant data are presented in table 4.7. In the 1970s, rapid labour outmigration combined with a high rate of agricultural growth to reduce agricultural underemployment in the countries of Near East-North Africa. The picture was similar, though less dramatic, in Latin America-Caribbean. Labour outmigration was slow and the rate of growth of the agricultural labour force was correspondingly high in both sub-Saharan Africa and Asia-Pacific. But the countries in Asia-Pacific achieved quite high rates of agricultural growth and this played the major role in reducing agricultural underemployment. In sub-Saharan Africa, agricultural growth barely exceeded the growth of the labour force and the fact that underemployment hardly declined at all must be attributed in large measure to agricultural stagnation.

In the 1980s, the growth of the agricultural labour force accelerated significantly only in Near East-North Africa. However, this was due not so much to a slowdown in outmigration but to demographic factors which led to an acceleration in the growth of the total labour force. Agricultural growth, on the other hand,

Table 4.7. Growth of labour force and GDP in agriculture (median)

Region	Labour force		GDP	
	1969-79	1979-85	1969-79	1979-85
SSA	1.5 (39)	1.5 (39)	1.6 (32)	0.0 (32)
AP	1.4 (15)	1.4 (15)	3.0 (14)	3.3 (14)
LAC	0.9 (21)	1.1 (21)	2.9 (20)	1.9 (20)
NENA	−0.3 (11)	0.4 (11)	4.4 (9)	2.1 (9)
DC	1.3 (86)	1.3 (86)	2.7 (75)	1.4 (75)

Note. Figures in parentheses indicate the number of countries for which data were available.
Source. Appendix tables II and IV.

Table 4.8. Growth of arable land area (percentages), median values

Region	1969-79	1979-85
SSA	0.81 (39)	0.20 (39)
AP	0.48 (15)	0.25 (15)
LAC	1.22 (21)	0.40 (21)
NENA	0.38 (11)	0.40 (11)
DC	0.82 (86)	0.26 (86)

Note. Figures in parentheses indicate the number of countries for which data were available.
Source. Appendix tables II and VII.

Table 4.9. Agricultural exports as a percentage of agricultural GDP, median values

Region	1969	1979	1985
SSA	27.1 (39)	18.6 (37)	17.6 (37)
AP	11.3 (14)	11.5 (14)	9.9 (14)
LAC	43.6 (19)	39.1 (19)	33.1 (19)
NENA	20.6 (11)	8.7 (10)	4.9 (8)
DC	26.5 (81)	19.3 (80)	17.3 (75)

Note. Figures in parentheses indicate the number of countries for which data were available.
Source. Appendix tables IX.

decelerated everywhere except in Asia-Pacific where it accelerated somewhat. Nevertheless, in Latin America-Caribbean and Near East-North Africa, the rate of growth of output was well above that of the labour force. But in sub-Saharan Africa, the rate of agricultural growth actually fell to zero.

The deceleration in the declining trend in agricultural underemployment in both Latin America-Caribbean and Near East-North Africa and the worsening conditions in sub-Saharan Africa particularly in the post-1979 period, therefore, are to be attributed primarily to the deteriorating production performance of agriculture rather than to a slackening of labour outmigration from agriculture. While we cannot go into a full investigation of the causes of agricultural decline, some relevant observations are obviously in order.

We have already noted that, in terms of land-augmenting technological change, sub-Saharan Africa performed poorly in both periods, while Latin America-Caribbean suffered a setback in the 1980s. Moreover, the rate of expansion of arable land area decelerated in all regions except Near East-North Africa in the 1980s, as can be seen from the data presented in table 4.8. For the countries in sub-Saharan Africa and Latin America-Caribbean, there is perhaps another important explanatory factor. Agricultural production in these regions has been, and remains, export-oriented to a significant degree, as can be seen from table 4.9. It is known that the global recession in the post-1979 period adversely affected both the demand for, and the prices of, agricultural exports from developing countries. In all likelihood, therefore, the deceleration in agricultural growth in sub-Saharan Africa and Latin America-Caribbean is in part attributable to stagnating exports. When viewed in conjunction with the fact that the rate of labour outmigration from agriculture was not appreciably affected by the decelerating industrial growth, it seems clear that the effect of the global recession on agricultural underemployment was transmitted directly rather than through the industrial sector.

Finally, there is the question of state policy. It is widely held that state policy has generally been discriminatory towards agriculture in developing countries.[18] This view, however, has been based essentially on studies of price policy. Clearly, if price policy has been discriminatory towards agriculture in most countries, then this has led to quite diverse results in different countries. The fact is that price policy, by itself, does not explain much; it needs to be viewed together with investment policy – a much neglected area of study. Be that as it may, it clearly is the case that agriculture was neglected in sub-Saharan Africa and, to an extent, also in Latin America-Caribbean. The main effect of this neglect, however, was on underemployment in agriculture and not on labour outmigration from agriculture.

4.4 Labour outmigration and agricultural underemployment: Some further obvservations

The thrust of the argument so far has been that the worsening employment conditions in agriculture in many countries can be attributed largely to the worsening production performance of agriculture. To say this, however, is not to say that labour outmigration played no role in the process. We have already noted that the growth of arable land area has been decelerating in all regions except Near East-North Africa. On the one hand, this has been one of the factors which influenced the overall production performance of agriculture. On the other hand, this has meant that arable land per worker has been declining at an increasing rate in most countries, particularly in sub-Saharan Africa and Asia-Pacific, even though the rate of labour outmigration has been relatively steady throughout the period.

A declining land-worker ratio, as stated earlier, has a negative impact on employment per worker and, therefore, increases underemployment. In fact, this is tautologically true in situations where availability of labour is not a binding constraint on the growth of land-productivity. The regression results presented in table 4.10, therefore, should not surprise anyone. They merely serve to emphasise the points that a declining land-worker ratio posed a real obstacle to reducing underemployment in agriculture and that the importance of this obstacle increased over time. What was needed was an accelerating rather than a steady pace of labour outmigration and in this sense outmigration had been increasingly inadequate in most countries, particularly in sub-Saharan Africa and Asia-Pacific.

[18] See, for example, World Bank (1983).

Table 4.10. Regression results, dependent variable: GLPR

Period	Regression type	Constant	GLDW	R^2	n
1969-79	Linear	1.65 (5.60)	0.57* (3.77)	0.16	75
1979-85	Linear	1.18 (3.54)	1.41* (5.53)	0.30	75

Note. GLPR = rate of growth of output per worker. GLDW = rate of growth of arable land per agricultural worker.
* Significant at 1 per cent.

It is also worth pointing out that a declining land-worker ratio can, by itself, generate conditions which constrain growth and technological change in agriculture. For growing underemployment and poverty often create conditions for the emergence of a variety of tenancy relationships and for personalisation of labour relationships.[19] Such relationships, in turn, discourage investment and adoption of new technologies. On the other hand, rapid labour outmigration and the consequent rise in the land-worker ratio can help accelerate agricultural growth. In the first place, this can lead to a dismantling, or at least a weakening, of the production relations which act as constraints on growth. Secondly, labour outmigration, when this does not merely mean an urbanisation of surplus labour, is usually associated with flow of remittances into agriculture and this can boost the investment capacity of agriculturists. Thirdly, rising output per worker in agriculture usually implies a steady growth of the marketed surplus of agricultural output and this boosts investment in the modern sector and strengthens agriculture-industry linkages.

These observations help place the issue of labour absorption in agriculture in a proper perspective. It has been observed, in the context of Asia, that there is considerable scope for increasing labour use per unit of land without reducing labour productivity.[20] Promotion of labour absorption under these conditions is obviously a desirable objective. But this cannot realistically be viewed as a mechanism for reducing the rate of labour outmigration from agriculture or as a substitute for employment promotion in non-agriculture and particularly in the modern sector. If land per worker continues to fall, increasing labour use per unit of land can have only a small impact on agricultural underemployment. Thus the decline in the land/worker ratio needs to be halted and that means accelerating the rate of outmigration from agriculture.

4.5 Conclusions

Agricultural underemployment is a very serious problem in developing countries; the indications are that around 50 per cent of the agricultural workers may

[19] See Bhaduri (1983); Ghose, ed. (1983), and Bharadwaj (1989).
[20] cf. ILO/ARTEP (1980a, 1980b), Gooneratne (1982) and Booth and Sundrum (1985).

be underemployed. The problem is most serious in sub-Saharan Africa and least serious in Near East-North Africa; it is of similar magnitudes in Asia-Pacific and Latin America-Caribbean.

In the seventies, agricultural underemployment declined in all regions except sub-Saharan Africa. In the eighties, it increased in sub-Saharan Africa and very probably also in Latin America-Caribbean. Although underemployment appears to have declined in Near East-North Africa in the eighties, the decline was rather small. Only in the countries of Asia-Pacific did it continue to decline at roughly the same rate as in the seventies.

The trends in underemployment are explained primarily by the growth performance of agriculture. In sub-Saharan Africa, agricultural growth was slow in the seventies and fell to zero in the eighties. In Latin America-Caribbean and Near East-North Africa, agricultural growth decelerated significantly. In Asia-Pacific, on the other hand, agricultural growth was slightly higher in the eighties than in the seventies. Technological developments were consistent with these patterns.

Even though the pace of labour outmigration from agriculture was similar in all regions in the two periods, the agricultural labour force grew at a higher rate in the eighties than in the seventies in Near East-North Africa mainly because of demographic changes. In all other regions, the rate of growth of the agricultural labour force was similar in the two periods. However, even this meant, particularly in sub-Saharan Africa and Asia-Pacific, a rapidly declining land/worker ratio, which posed real obstacles to reducing underemployment.

In this context, the pace of labour outmigration needs to be accelerated rather than decelerated and promotion of labour absorption in agriculture, while necessary, cannot serve as a substitute for employment promotion in non-agriculture.

5

Issues of development policy

5.1 An overview

The problem of employment confronting the developing countries has no parallel in history. These countries embarked on what Kuznets has called modern economic growth at a point in time when they not only had massive quantities of surplus labour in agriculture but also faced rapid and accelerating demographic expansion. The available evidence suggests that even in the late seventies around 50 per cent of the labour force in agriculture could be said to be underemployed. During the period considered in this study, the labour force was growing at an average annual rate of 2.5 per cent and the growth was still accelerating.

There were some fortunate circumstances. Since the stage of demographic transition corresponded to the level of development, the rate of growth of the labour force tended to be higher in relatively more developed countries. In other words, the challenge was bigger for countries which, in principle, had a greater capacity to cope. There is, moreover, no evidence to suggest that demographic growth hindered economic growth. In part, this is a reflection of the fact that governments were more concerned with economic growth than with social development. More fundamentally, however, economic growth was determined by factors which were quite independent of demographic variables.

In the seventies, the developing countries as a group faced up to the challenge reasonably well.[1] The rate of growth of wage-employment generally exceeded that of the labour force in the non-agricultural sector, the pace of labour outmigration from agriculture was in conformity with the early experience of the industrialised countries and underemployment in agriculture generally declined. The exceptions to the rule were to be found among the least developed countries of sub-Saharan Africa.

In the eighties, however, a large number of countries faced economic difficulties the reasons for which are by now well-known: the recession in industrialised countries, the decline of prices of primary commodities in international markets and the debt crisis. The ensuing economic stagnation or decline gave some unexpected twists to the process of structural transformation. Non-agricultural wage employment as well as real wages declined in many countries, but the pace of outmigration from agriculture hardly slowed and agricultural underemployment

[1] The same was true, according to Gregory, in the sixties. See Gregory (1980).

either remained unchanged or worsened. The exceptions to the rule this time were to be found among the countries of Asia-Pacific which remained relatively unaffected by the changes in the global economy.

The contrasting experiences of the two periods already suggest important inadequacies in the analytical schemes which have so far dominated thinking on development. An analysis of the details shows that the demand-side factors such as Engel's Law and structure of exports and the supply-side factors such as intersectoral differentials in productivity growth, emphasised in the Clark-Kuznets analytical tradition, were not of much importance in bringing about intersectoral shifts in employment. The actual process, in the seventies, resembled much more that charted by the dual economy models – capital accumulation in the modern sector was a major driving force and intersectoral shifts were generally associated with shifts in the mode of employment. In the eighties, however, the modern sector in many countries faced a crisis but this had little impact on the pace of labour outmigration from agriculture. Intersectoral shifts of the labour force, in this period, did not signify genuine structural change. Clearly the dual economy models, too, cannot adequately account for the intersectoral shifts of the labour force in developing economies.

As a matter of fact, the non-agricultural sector cannot be regarded as equivalent to the modern sector of the dual economy models, though such an equivalence has often been supposed in development literature. In fact the non-agricultural sector in most countries was found to be highly dualistic; it typically included a large traditional segment. The agricultural sector, on the other hand, can in general be considered as traditional, even though wage-labour is quite significant in many countries. Wage-employment in agriculture rarely resembles that in the modern part of non-agriculture. Labour-hiring farms are typically not modern business enterprises, the labourers are often traditional peasants and wage-employment is usually irregular. Underemployment exists as much among wage-labourers as among traditional peasants.

Given this context, we should ideally consider three sectors – traditional agriculture, traditional non-agriculture and modern non-agriculture – and postulate two types of labour flows: from agriculture to traditional non-agriculture and from traditional non-agriculture to modern non-agriculture. Evidence shows that there need be no necessary relation between these two types of labour flows. In the seventies, rapid growth of the modern non-agricultural sector had a stimulating effect on labour outmigration from agriculture, i.e. the two types of labour flows were positively related. In the eighties, however, stagnation of the modern non-agricultural sector did not significantly discourage labour outmigration from agriculture, i.e. even though the labour flow from traditional non-agriculture to modern non-agriculture was halted or even reversed, the labour flow from agriculture to traditional non-agriculture was not appreciably affected.

In fact, our analysis shows that, in the eighties, labour outmigration from agriculture was explained primarily by the "push factors", i.e. those associated with growing scarcity of arable land in relation to the labour force. What appears to be the case in general is that the push factors engender a certain base rate of labour outmigration from agriculture into traditional non-agriculture, irrespective of the labour flow from traditional non-agriculture to modern non-agriculture. When the latter type of labour flow is significant (because of rapid growth of modern non-agriculture), it has the effect of raising the rate of labour outmigration from agriculture above the base rate.

These observations lead to four conclusions. First, the share of agriculture in the labour force will in general tend to decline irrespective of the state of the modern non-agricultural sector. But this should obviously not be taken as an indicator of structural change; the true indicator of structural change is the share of the modern sector in total employment. Capital accumulation in the modern sector acts as a driving force in bringing about intersectoral shifts in employment only when the former is adequate in the sense that it enables wage-employment to grow at a rate significantly higher than that of the labour force. Even the dual economy models, therefore, fail to capture an important aspect of the process because they ignore both push factors and dualism within the non-agricultural sector.[2]

Secondly, the share of the traditional non-agricultural sector in the labour force will in general tend to vary inversely with that of modern non-agriculture. Such a tendency has been quite widely noted, particularly in the context of the Latin America-Caribbean region.[3] We need only add that traditional non-agriculture in this case includes both the rural non-farm sector and the urban informal sector.

Thirdly, the existence of a traditional non-agricultural sector is a logical consequence of the existence of surplus labour in agriculture. The importance of the traditional non-agricultural sector will begin to be eroded only when surplus labour in agriculture has disappeared, provided, of course, that the growth of the modern sector can be sustained. This can only be visualised as a long-term prospect. This means that the sector cannot be ignored in devising economic policy at the present juncture. We shall return to the point below.

Fourthly, in such a context, the Harris-Todaro argument is of little relevance in understanding the regulatory mechanisms of labour outmigration from agriculture. To put it sharply, workers outmigrate from agriculture into traditional non-agriculture (and not to the modern sector) and they do so because their marginal contribution to family incomes in agriculture is zero (note that this does not imply that the marginal productivity of labour in agriculture is zero). The only condition that needs to be satisfied for migration to occur is that the average labour-income in traditional non-agriculture is positive. Differentials between the average wage in the modern non-agricultural sector and the average income in agriculture are of no relevance. The experience of many developing countries in the eighties demonstrates this; the sharp decline in the modern sector wage did little to stem outmigration from agriculture.

Two further points which emerge from our analysis are also worth noting. We have viewed output growth in the manufacturing sector as the best indicator of the growth of the modern sector. The reason is not that the manufacturing sector accounts for the bulk of wage employment in non-agriculture. In fact, typically, only a small proportion of the non-agricultural wage employment is in manufacturing; the bulk of wage employment in developing countries is actually in services. But the growth

[2] It should be said, however, that Lewis, in his more recent writings, emphasises both push factors and dualism within the non-agricultural sector. See Lewis (1979).

[3] See Garcia and Tokman (1984) and Tokman (1988).

of wage employment in non-agriculture depends crucially on the growth of manufacturing output. Thus it can be said that accumulation occurs primarily in manufacturing and this makes it the leading sector. The fact that the manufacturing sector accounts for only a small proportion of non-agricultural wage employment suggests that the techniques used are indeed capital-intensive. This is hardly surprising in view of the fact that capital goods are typically imported from industrialised countries. It is important to point out, however, that the manufacturing sector has not become less and less labour-intensive over time.

The other point is about underemployment in agriculture. Rare was the case where agricultural underemployment declined solely because of labour outmigration from agriculture; agricultural growth was a major determinant of the trends in underemployment. The point is best illustrated by the contrasting experiences of Asia-Pacific and sub-Saharan Africa. The two regions had similar rates of growth of the agricultural labour force, and arable land per worker was slightly higher in sub-Saharan Africa. But agricultural underemployment declined in Asia-Pacific throughout the period while, in sub-Saharan Africa, it changed little in the seventies and increased in the eighties. The reason is simple. If labour outmigration from agriculture is not rapid enough to increase the arable land per worker, then underemployment can decline only if technological change increases both labour requirements and yields per unit of land. The evidence shows that land-augmenting technological change was quite impressive in Asia-Pacific but was negligible in sub-Saharan Africa. And, although we have not analysed this, it is known that technological change in agriculture is very dependent on state investment.[4] It is worth recording these facts because the dual economy models have created the impression, and this has undoubtedly influenced economic policy in developing countries, that agricultural underemployment declines as a result of labour outmigration, i.e. that agricultural employment need not be a direct object of development policy.[5]

5.2 Some issues of macro-economic policy

With respect to employment creation, developing countries, particularly the low-income developing countries, will be facing an even bigger challenge in the next decade or two than they did in the past. The rate of growth of the labour force will accelerate in poorer countries and will remain stable in high-income countries; it is likely to be around 2.8 per cent in a majority of the countries. The vigorous implementation of family planning programmes is an urgent necessity. But it must be recognised that even if this is done, the rate of growth of the labour force in the next two decades will not be appreciably affected.

[4] It has been pointed out, however, that "a substantial rate of increase in agricultural production can be achieved largely through the more effective use of resources already committed to the agricultural sector and with only modest requirements for the critical resources of high opportunity cost ...", Johnston and Mellor (1961), p. 570. The implication is that it is not so much the volume as the pattern of public investment which is of crucial importance.

[5] Lewis had worried about the effects of a rise in labour productivity in agriculture on the wage rate in the modern sector. However, in a more recent paper (Lewis, 1974) he emphasises the importance of increasing labour productivity in agriculture and blames others for misinterpreting his ideas.

At the same time, the growth of arable land area will, in all probability, decelerate quite rapidly. Again, it is possible to think of measures which might help expand arable land area, but it is difficult to visualise situations where the process of deceleration can be arrested or reversed.

The two facts together imply that, irrespective of the growth policies adopted, the pace of outmigration from agriculture is likely to accelerate, particularly in the low-income countries. Growth policies, in this context, will determine two things: the direction of change in agricultural underemployment and the relative size of the traditional non-agricultural sector.

But can growth be relied upon to cope with the employment problem or should employment be an independent objective of macro-economic policy?[6] The answer seems to be that there is no real choice here. The available evidence suggests that economic growth in contemporary developing countries can be quite rapid and that, if this is so, employment conditions do improve. If, on the other hand, growth falters, then this indicates a crisis of accumulation, i.e. a lack of resources in the hands of investors including the state. In that case, it is hardly feasible to promote employment.

Having said this, it should be acknowledged that patterns of growth matter in ensuring sustainability of the growth process and, therefore, in employment generation. In particular, maintenance of a proportionality relationship between growth of agriculture and that of the modern sector is of critical importance.[7] There is no doubt that such a relationship was not maintained in the past in many countries of sub-Saharan Africa and Latin America-Caribbean, and that this has contributed to the present crisis.[8] This is one aspect that should receive adequate attention in future development strategies.

There clearly is a serious crisis of accumulation in many developing countries today. Whether and how the crisis may be overcome is a large question that cannot be answered here; issues relating to the problems of external debt and international capital flows are involved. We must simply assume that the crisis of accumulation will be overcome and proceed to suggest some desirable objectives – desirable in terms of employment effects – for macro-economic policy in the future. Given the premises thrown up by our analysis, attention is focused on policy objectives in three sectors – modern non-agriculture, agriculture and traditional non-agriculture. In what follows, the three sectors are viewed as complementary rather than as competitive; simultaneous growth of all of them is needed if the emerging employment problem is to be effectively tackled. The problem is one of achieving a proper balance among the sectors. But the definition of "proper balance" can only be country-specific. The

[6] The question received some attention from economists. See Morawetz (1975) for a review of the arguments.

[7] This is an old, though neglected, argument. See, for example, Lewis (1958), Kaldor (1980), Kuznets (1965), Kalecki (1976) and Mellor and Johnston (1984). The basic arguments are briefly stated at a later point in the text.

[8] Our analysis has shown that the neglect of agriculture has led to growing underemployment in agriculture in spite of significant outmigration. It is also arguable that the neglect of agriculture, particularly of the food sector, also lies at the root of the industrialisation crisis, but the argument cannot be pursued here.

discussion below does no more than dwell on some issues of general relevance; a detailed discussion of the country- or region-specific policies falls outside the scope of this study.

Creation of regular wage employment in the modern non-agricultural sector at a rate higher than the rate of growth of the non-agricultural labour force remains both necessary and desirable. It is necessary because, if this does not happen, surplus labour will continue to accumulate in the traditional non-agricultural sector (i.e. in the rural non-farm sector and in the urban informal sector), and the social consequences can be very serious. It is desirable because development, as so often observed, is really about transferring labour from low productivity self-employment to high-productivity wage employment.

The objective is not easy to achieve, since the rate of growth of non-agricultural labour force is high and is likely to accelerate in the near future. But it is not impossible to achieve; it was actually achieved in many developing countries in the past. We have argued that the growth of wage employment in modern non-agriculture is primarily a function of the growth of manufacturing output. It follows that rapid growth of manufacturing output should be a major objective of macro-economic policy. This is an old proposition which needs to be emphasised only because, in current debates on structural adjustment policies, a tendency to de-emphasise the role of industrialisation in development can be discerned.

It is useful to examine, in this context, the currently dominant view that developing countries should pursue an export-oriented rather than an import-substitutive industrialisation strategy. Two arguments are usually advanced in support of this view. First, it is argued that import-substitutive industrialisation is necessarily associated with protectionist policies and that this leads to long-run inefficiency; infant industries do not show a great propensity to grow into adulthood.[9] This argument is really about management of the transition from a protectionist to a liberal trade regime in the course of industrialisation and is not against import-substitutive industrialisation as such.[10] It has not been shown that developing countries today can industrialise without protectionist policies; indeed, the problem is not merely one of building industries but also of building a domestic entrepreneurial class. There is considerable evidence to suggest that a prior phase of import-substitutive industrialisation is needed before export-oriented industrialisation can begin.[11]

Secondly, it has been argued that export-oriented industrialisation is more employment-intensive than import-substitutive industrialisation.[12] Theoretically, the argument derives from the idea that a country has a comparative advantage in those commodities whose production involves greater use of the relatively abundant factor

[9] See Balassa (1980), Krueger (1980), World Bank (1987) and Pack (1988).

[10] For arguments (other than the infant industry argument) in support of import-substitution, see Prebisch (1959), Singer (1950, 1987) and Mahalanobis (1953). For an analysis of evidence on the Prebisch-Singer hypothesis, see Grilli and Yang (1988). A much neglected argument, developed in Lewis' open economy model (Lewis, 1954), is also relevant in this context. For a reformulation, see Bardhan (1982).

[11] See, for example, Lee, ed. (1981).

[12] See Krueger (1983, 1988).

available to it.[13] Since the developing countries have abundance of labour and shortage of capital, their comparative advantage should lie, according to this view, in labour-intensive commodities. There is some evidence to show that exports of developing countries are more labour-intensive than domestically consumed products; but the evidence is by no means conclusive. Besides, if we are concerned not with the level of industrial employment at a point of time but with the growth of industrial employment, then it has to be shown that, for any given rate of industrial growth, an export-oriented strategy is associated with a higher rate of growth of employment than an import-substitution strategy. Aggregative indicators such as employment elasticities of manufacturing output do not seem to support such a proposition; for example, the elasticity turns out to be roughly the same in the case of the Republic of Korea as in the case of India.

Our argument is that employment effects of manufacturing growth should not be seen in terms of employment generated in the manufacturing sector alone; the rate of growth of manufacturing output determines the rate of growth of wage employment in the entire modern non-agricultural sector.[14] Optimisation of the rate of growth of manufacturing output is what is important and policies designed to promote employment-intensity of manufacturing output are not necessarily the most appropriate for the purpose. In some cases, where the dominant constraint on manufacturing growth is foreign exchange availability, export expansion may indeed be required. But often the dominant constraint is the slow growth of food output and it may well be appropriate to concentrate on capital and intermediate goods industries geared to the needs of agricultural development. Generalisations, in these matters, are illegitimate.[15]

Having said this, however, we cannot fail to recognise that the growth of manufacturing exports is an essential concomitant of the growth of manufacturing output. Manufacturing growth (irrespective of whether this is import-substitutive or export-oriented) in developing countries, under modern conditions, is necessarily import-intensive. It is also necessary, as will be argued below, to progressively replace agricultural exports by manufacturing exports. Thus once the domestic constraints (savings and food) are overcome, the rate of growth of manufacturing output becomes dependent on the rate of growth of imports and, therefore, on the rate of growth of exports. And given the need to progressively replace agricultural exports, the rate of growth of manufacturing exports assumes critical importance. This, rather than the need to increase the employment elasticity of manufacturing output, is the really relevant argument for export-oriented industrialisation.

We can now turn to policies relating to agriculture. It has long been an implicit presumption of development theory that labour outmigration from agriculture,

[13] The theoretical argument is due to Stolper and Samuelson (1941).

[14] Indeed, the Krueger view can be justifiably criticised for its exclusive focus on tradeables.

[15] A brief comment on a related issue may be usefully added at this point. It has long been supposed that wage rates in modern non-agriculture are important determinants of the pace of labour outmigration from agriculture. Our analysis has shown this to be a wrong presumption. Regulating the pace of labour outmigration from agriculture, therefore, should not be seen as an objective of wage policy in the manufacturing sector.

induced by industrial growth, will by itself lead to a decline of underemployment in agriculture. This view happens to be consistent with another old idea that, in initial stages of development, it is appropriate to squeeze a surplus out of agriculture for financing industrialisation.[16] Development experience has shown that neither of these ideas is of much relevance.

Recent research has shown that the direction of the flow of resources between agriculture and industry depends very much on the trends in labour productivity in agriculture. A growing agriculture provides resources for industrialisation independently of any efforts by the government to extract them. On the other hand, attempts to extract resources from a stagnant agriculture, through such means as taxation or manipulation of intersectoral terms of trade, tend to be self-defeating.[17]

More importantly, our analysis has shown that while labour outmigration from agriculture can be influenced by the growth of manufacturing output, it is unrealistic to suppose that labour outmigration alone can reduce underemployment in agriculture; rapid growth of agricultural output is also needed. It is worth pointing out, furthermore, that although agricultural growth is being viewed here as necessary for reducing underemployment, agricultural growth can also be regarded as important for another reason. Agriculture is the supplier of food – the basic wage-good in developing countries – and employment growth in modern non-agriculture cannot really be sustained without adequate agricultural growth. The reason is that the income elasticity of demand for food is high in developing countries and growth of wage employment is necessarily accompanied by growth in the demand for food in per capita terms. If the production and marketed surplus of food do not grow correspondingly, then either food imports will have to rise or the real wage rate in the modern non-agricultural sector will have to decline or the growth of non-agricultural employment itself will have to slow down. None of these options can be regarded as acceptable. There is also a further aspect to consider. The marketed surplus of food also constitutes a major source of demand for domestic manufactures. In other words, agricultural growth is very much linked to the expansion of the domestic market, which is necessary for sustained industrialisation.

The view that agricultural growth needs to be accorded priority has featured prominently in current debates on structural adjustment. Unfortunately, however, there has been a tendency to argue *(a)* that the export sector in agriculture should receive priority and *(b)* that appropriate price policies alone can stimulate agricultural growth.[18] Given the conditions in the global markets, there are reasons to believe that efforts by developing countries to increase agricultural exports will depress world prices and, in many cases, will reduce total export revenues. Meanwhile, the need for food imports will rise for the reasons explained above.[19] As for the relevance of price policies, two points need to be made. First, there is in fact little evidence to suggest that intersectoral terms of trade influence agricultural production to any significant

[16] For a detailed discussion, see Timmer (1988).

[17] Two interesting recent studies on the subject are Ishikawa (1988) and Karshenas (1989).

[18] See, for example, World Bank (1983, 1986).

[19] For a discussion on the issue of comparative advantage in food production, see Mellor and Johnston (1984).

extent.[20] An important reason is that intersectoral terms of trade are not good indicators of intersectoral resource flows. Even when intersectoral terms of trade are biased against agriculture, investment flows into agriculture may be more than compensatory. Indeed, government investment (including the provision of credit) is usually of crucial importance in determining the trends in output and technology in agriculture.[21] Adverse terms of trade, moreover, restrict income growth only for surplus agricultural producers but have beneficial effects on the level of living of the poorer agricultural households. At any rate, it is not at all certain that a combination of favourable terms of trade and low government investment can produce better results in terms of growth and technological change in agriculture than a combination of adverse terms of trade and high government investment. What is certainly undesirable is a combination of adverse terms of trade and low government investment.

On the other hand, secondly, there is a good deal of evidence to suggest that relative crop prices have a significant effect on the cropping pattern; the allocation of the existing resources to different crops within agriculture is responsive to relative crop prices.[22] It is in changing the cropping pattern that price policy can be regarded as an appropriate instrument. If, for instance, an emphasis on food crops (and a corresponding de-emphasis on export crops) is intended, increasing the relative prices of food crops vis-à-vis export crops would be an appropriate policy measure.

This indeed is a policy that we would recommend. For, in our view, the main objective of policies towards agriculture should be to stimulate the rate of growth of food production. This will not only serve to reduce underemployment in agriculture but will also help sustain the growth of the modern non-agricultural sector. There can be two arguments against this view. First, a de-emphasis on export crops may reduce foreign exchange availability. But it should be clear from our earlier arguments that this is by no means certain; a reduction in supply in the world market may well induce price increases so that export revenues need not decline. Moreover, since the policy is likely to lead to a progressive decline in the need for food imports, even a small decline in export revenues cannot do much harm. Secondly, it can be argued that, given the domestic cost conditions and international relative prices, it is better for some developing countries to import food than to produce it. Even if this argument is valid in the short run, however, it is unlikely to hold in the long run. International food prices are presently kept artificially low by the policies of massive subsidisation of agriculture in developed countries and these policies are very likely to be gradually abandoned. Furthermore, cost conditions in developing countries will alter if technological change can be brought about.

Past experience tells us that agricultural growth is dependent primarily on land-augmenting technological change. And this requires government investment. Without substantial government investment in irrigation development, research, extension, credit supply, etc., land-augmenting technological change is unlikely to

[20] See Krishna (1984) and Ghose (1987).
[21] On this point, cf. Timmer (1988).
[22] See Scandizzo and Bruce (1980).

occur (or will occur only over a very long period). In terms of policy, therefore, the primary emphasis must be on government investment.

The emphasis on land-augmenting technological change should be noted; the desired objective is to increase labour use per unit of land and labour productivity simultaneously. Promoting labour absorption in agriculture is important, though, it is worth repeating, this should not be viewed as a means to reduce the pace of labour outmigration from agriculture. Given the seriousness of the problem of underemployment in agriculture, labour outmigration needs to be stimulated, not reduced. The objective is not to retain the entire incremental labour force in agriculture but to make better use of the labour of those who stay in agriculture.

Our analysis has suggested that the traditional non-agricultural sector (incorporating rural non-farm and urban informal sectors) will remain important in terms of its share of the labour force for a long time to come; this will be the case even if the growth of agriculture and manufacturing is satisfactory. Indeed, if agricultural growth is sustained, the traditional non-agricultural sector may even become the major reservoir of surplus labour. The existence of the sector, therefore, cannot be ignored in policy analysis.

It is known that activities or enterprises in this sector are of two types.[23] The bulk of them represent no more than survival strategies of the poor and the underemployed. But others can potentially contribute both to growth and to employment generation.

There are two critical questions to answer in this area. First, how does the output/capital ratio in this sector compare with that in the modern sector? Clearly, it is necessary that the output/capital ratio in this sector be at least equal to that in the modern sector; otherwise a diversion of resources to the sector, while increasing employment, will reduce growth. In fact, even if the above condition is satisfied, the economic surplus in the sector will still be lower (because employment is higher) than in the modern sector unless the labour-incomes are lower in the former. But it is perhaps arguable that lower labour-incomes in such a context are acceptable.

While a certain amount of empirical research on the question has been done, this has not produced a conclusive answer as yet.[24] Some recent studies suggest that the answer to the question should be sought not on an aggregative level but on a product-by-product basis.[25] For a particular product, as a general rule, household and small-scale enterprises tend to have lower output/capital ratios than medium or large-scale enterprises. But the definition of "medium or large-scale" is relative; it varies a great deal across products. Thus there may well be products which are more efficiently produced in the traditional non-agricultural sector. But the identification of these will require extensive country-specific research.

[23] See Tokman (1978), Mukhopadhyay and Lim, eds. (1985), Islam, ed. (1987), Chuta and Sethuraman, eds. (1984) and Saith (1989b).

[24] See Steed and Takagi (1983) for a brief review of the evidence and references. See, in addition, Sen (1968), Bhalla (1965) and Dandekar and Rath (1971).

[25] See Little (1987) and Little, Mazumdar and Page (1987).

The second question relates to demand. It is often argued that supply-side policies such as measures to upgrade skills and technology, provision of credit, etc., can help integrate this sector into the mainstream economy. However, it is difficult to see how these policies can help unless the growth of demand for the sector's products can be ensured.[26] Although the problem of demand has often been noted in the literature, policy recommendations have rarely gone beyond vague suggestions about development of co-operative marketing, exploitation of the market associated with tourism, etc.

If in an economy the agricultural and the modern sectors are growing rapidly, demand is unlikely to be a constraint on the growth of the traditional non-agricultural sector. But this is precisely the situation where there is little need to worry about the potential of the traditional non-agricultural sector for employment generation. It is when the growth of the other sectors is slow that the role of the traditional non-agricultural sector is highlighted; but this is also when a decent growth of demand for the sector's products can be ensured only by expanding the sector's share of the overall market, i.e. by constraining the growth of demand for the modern sector's products.

In this context, the question that needs to be asked is: are there goods and services, with potential for demand growth, which can be reserved exclusively for production in the traditional non-agricultural sector? There are two main reasons why this question is of central importance.[27] First, if the traditional non-agricultural sector has to compete with the modern non-agricultural sector in both input and output markets, the probability of the former's survival (except as a marginal sector) is not high. Secondly, diversion of investable resources from agriculture and modern non-agriculture can be justified, in the context of a growth strategy, only if this leads to a self-sustained process of growth of the traditional non-agricultural sector. Otherwise, the diversion will amount to subsidisation and will stretch over an indefinite time. The economy will be forced on to a lower growth path as a consequence; the workers in traditional non-agriculture will in effect be trading some increases in incomes now against future prospects of transferring to modern non-agriculture.

It must be said, however, that where the output/capital ratios in traditional non-agriculture compare favourably with those in modern non-agriculture and the growth of demand for the output of the traditional non-agricultural sector can be ensured, there is a considerable advantage in diverting resources to traditional non-agriculture. Not only does this help to accelerate the growth of productive employment outside agriculture but it also reduces pressures on the balance of payments. Growth of the modern sector is much more import-intensive than that of traditional non-agriculture.

[26] For a good discussion of the issues involved, see Saith (1989b).

[27] The idea of providing a protected market for the traditional non-agricultural sector was proposed by Mahalanobis (1958) and the Government of India attempted to implement it. For a penetrating evaluation of the results, see Dandekar and Rath (1971).

Indeed, a severe foreign exchange constraint may force some governments to concentrate on the development of the traditional non-agricultural sector in any case. But it must be recognised that unless simultaneous efforts are made to ensure efficiency of capital-use in the sector and to stimulate demand for the sector's output, it will constitute no more than a survival strategy at the national level.

Bibliography

Bairoch, P. 1975. *The economic development of the Third World since 1900,* translated by Lady Cynthia Postan (London, Methuen and Co.).

– – –; Limbor, J.M. 1968. "Changes in the industrial distribution of the world labour force", in *International Labour Review,* Oct. 1968.

Balassa, B. 1979. "A stages approach to comparative advantage", in Adelman, I. (ed.), *Economic growth and resources,* Vol. 4 (London, Macmillan).

– – –. 1980. *The process of industrial development and alternative development strategies,* World Bank Staff Working Paper No. 438 (Washington, DC).

Bardhan, P.K. 1982. "Unequal exchange in a Lewis-type world", in Gersovitz et al. 1982.

Bell, C. 1987. "Development economics", in Eatwell et al. (eds.), 1987.

Bhaduri, A. 1983. *Economic structure of backward agriculture* (London, Academic Press).

Bhalla, A.S. 1965. "Choosing techniques: hand-pounding v. machine milling of rice", in *Oxford Economic Papers,* Vol. 17, No. 1.

Bharadwaj, K. 1989. *The formation of rural labour markets: An analysis with special reference to Asia* (Geneva, ILO; mimeographed World Employment Programme Working Paper; restricted).

Binswanger, H.P.; Rosenzweig, M.R. 1984. *Contractual arrangements, employment and wages in rural labour markets: A critical review* (New Haven, York University Press).

Birdsall, N. 1988. "Economic approaches to population growth and development", in Chenery and Srinivasan (eds.), 1988.

Bliss, C.J.; Stern, N. 1978. "Productivity, wages and nutrition: Part I: The theory, Part II: Some observations", in *Journal of Development Economics,* Vol. 5, No. 4.

Bloom, D.; Freeman, R. 1987. "Population growth, labour supply and employment in developing countries", in Johnson and Lee (eds.), 1987.

Booth, A.; Sundrum, R.M. 1985. *Labour absorption in agriculture* (New Delhi, Oxford University Press).

Boserup, E. 1965. *The conditions of agricultural growth* (London, Allen and Unwin).

– – –. 1981. *Population and technological change* (Chicago, Chicago University Press).

Chenery, H.; Syrquin, M. 1975. *Patterns of development, 1950-1970* (London, Oxford University Press).

– – –; Robinson, S.; Syrquin, M. 1986. *Industrialisation and growth* (New York, Oxford University Press).

---; Srinivasan, N. (eds.). 1988. *Handbook of development economics* (Amsterdam, North Holland).

Chuta, E.; Sethuraman, S.V. (eds.). 1984. *Rural small-scale industries and employment in Africa and Asia* (Geneva, ILO).

Clark, C. 1957. *The conditions of economic progress,* 3rd ed. (London, Macmillan).

Dandekar, V.M.; Rath, N. 1971. *Poverty in India* (Poona, Indian School of Political Economy).

de Janvry, A.; Sadoulet, E. 1986. *Rural labour in Latin America* (Geneva, ILO; mimeographed World Employment Programme Working Paper; restricted).

Dobb, M. 1963. *Economic growth and under-developed countries* (London, Lawrence and Wishart).

Eatwell, J.; Milgate, M.; Newman, P. (eds.). 1988. *The new Palgrave dictionary of economics* (London, Macmillan).

Eckaus, R.S. 1955. "The factor-proportions problem in under-developed areas", in *American Economic Review,* Vol. 45, No. 4.

Eicher, C.K.; Staatz, J.M. (eds.). 1984. *Agricultural development in the Third World* (Baltimore, Johns Hopkins University Press).

Fei, J.C.H.; Ranis, G. 1964. *Development of the labour surplus economy* (Homewood, Illinois, Irwin).

Garcia, N.; Tokman, V. 1984. "Changes in employment and the crisis", in *CEPAL Review,* 24 Dec.

Gerschenkron, A. 1962. *Economic backwardness in historical perspective* (Cambridge, Massachusetts, Harvard University Press).

Gersovitz, M.; Dias-Alejando, C.F., Ranis, G.; Rosensweig, M.R. (eds.). 1982. *The theory and experience of economic development: Essays in honour of Sir W. Arthur Lewis* (London, Allen and Unwin).

Ghai, D.; Radwan, S. (eds.). 1983. *Agrarian policies and rural poverty in Africa* (Geneva, International Labour Office).

Ghose, A.K. 1979. "Wages and employment in Indian agriculture", in *World Development,* Vol. 8, Nos. 5-6.

---. (ed.). 1983. *Agrarian reform in contemporary developing countries* (London, Croom-Helm).

---. 1987. *Agriculture-industry terms of trade and distributive shares in a developing economy* (Geneva, ILO; mimeographed World Employment Programme research working paper; restricted).

---. 1989. "Rural poverty and relative prices in India" in *Cambridge Journal of Economics,* Vol. 13, No. 2.

Gooneratne, W. (ed.). 1982. *Labour absorption in rice-based agriculture: Case studies from South-east Asia* (Bangkok, ILO/ARTEP).

Gregory, P. 1980. "An assessment of changes in employment conditions in less developed countries", in *Economic Development and Cultural Change,* Vol. 28, No. 4.

Grilli, E.R.; Yang, M.C. 1988. "Primary commodity prices, manufactured goods prices, and the terms of trade of developing countries: What the long run shows", in *The World Bank Economic Review,* Vol. 2, No. 1.

Harris, J.R.; Todaro, M.P. 1970. "Migration, unemployment and development: A two-sector analysis", in *American Economic Review,* Vol. 60, No. 1.

Horlacher, D.E.; Mackellar, F.L. 1988. "Population growth versus economic growth", in Salvatore (ed.), 1988.

Hayami, Y.; Ruttan, V. 1985. *Agricultural development: An international perspective* (Baltimore, Johns Hopkins University Press).

ILO. 1970. *Towards full employment: A programme for Colombia* (Geneva).

---. 1971. *Matching employment opportunities and expectations: A programme of action for Ceylon* (Geneva).

---. 1972. *Employment, incomes and equality: A strategy for increasing productive employment in Kenya* (Kenya).

ILO/ARTEP. 1980a. *Labour absorption in agriculture: The East Asian experience* (Bangkok).

---. 1980b. *Employment expansion in Asian agriculture: A comparative analysis of South Asian countries* (Bangkok).

ILO/JASPA. *The challenge of employment and basic needs in Africa* (Nairobi, Oxford University Press).

Ishikawa, S. 1988. "Patterns and processes of inter-sectoral resource flows: Comparison of cases in Asia", in Ranis and Schultz (eds.), 1988.

Islam, R. (ed.). 1987. *Rural industrialisation and employment in Asia* (New Delhi, ILO/ARTEP).

Jamal, V. 1986. "Rural-urban gap, income distribution and poverty in Africa", in ILO/JASPA, 1986.

---; J. Weeks. 1988. "The vanishing rural-urban gap in sub-Saharan Africa", *International Labour Review,* Vol. 127, No. 3.

Johnson, D.G.; Lee, R. (eds.). 1987. *Population growth and economic development: Issues and evidence* (Madison, Wisconsin University Press).

Johnston, B.F.; J.W. Mellor. 1961. "The role of agriculture in economic development", *American Economic Review,* Vol. 51, No. 4.

Kakwani, N. 1989. "On measuring undernutrition", in *Oxford Economic Papers,* Vol. 41, No. 3.

Kaldor, N. 1980. "What is wrong with economic theory?" in *Further essays in economic theory* (London, Duckworth).

Kalecki, M. 1976. *Essays on developing economies* (Brighton, Harvester Press).

Kanbur, R.; MacIntosh, J. 1987. "Dual economics", in Eatwell et al. (eds.), 1987.

Karshenas, M. 1989. *Intersectoral resource flows and development: Lessons of past experience* (Geneva, ILO; mimeographed World Employment Programme research working paper; restricted).

Kelley, A.C. 1988. "Economic consequences of population change in the Third World", in *Journal of Economic Literature,* Vol. 26, No. 4.

Khan, A.R.; Lee, E. (eds.). 1984. *Poverty in rural Asia* (Bangkok, ILO/ARTEP).

Krishna, Raj. 1984. "Price and technology policies", in Eicher and Staatz (eds.), 1984.

Krueger, A.O. 1980. "Trade policy as an input to development", in *American Economic Review,* Papers and Proceedings, Vol. 70, No. 2.

---. (ed.). 1983. *Trade and employment in developing countries* (Chicago, University of Chicago Press).

---. 1988. "The relationships between trade, employment and development", in Ranis, G. and Schultz, T.P. (eds.), 1988.

Kuznets, S. 1965. *Economic growth and structure: Selected essays* (London, Heinemann).

---. 1971. *Economic growth of nations: Total output and production structure* (Cambridge, Massachusetts, The Belknap Press).

Leamer, E. 1984. *Sources of international comparative advantage* (Cambridge, Massachusetts, MIT Press).

Lee, E. (ed.), 1981. *Export-led industrialisation and development* (Singapore, Maruzen Asia for ILO/ARTEP).

Lewis, W.A. 1954. "Economic development with unlimited supplies of labour", in *The Manchester School,* Vol. 22, No. 2.

---. 1958. "Unlimited labour: Further notes", in *The Manchester School,* Vol. 26, No. 1.

---. 1979. "The dual economy revisited", in *The Manchester School,* Vol. 47, No. 3.

Little, I.M.D. 1987. "Small manufacturing enterprises in developing countries", in *World Bank Economic Review,* Vol. 1, No. 2.

---; Mazumdar, D.; Page, J.M. 1987. *Small manufacturing enterprises: A comparative study of India and other countries* (New York, Oxford University Press).

Mahalanobis, P.C. 1953. "Some observations on the process of growth of national income", in *Sankhya,* Vol. 16, No. 1/2.

---. 1955. "The approach of operational research to planning in India", in *Sankhya,* Vol. 12, No. 4.

Mazumdar, D. 1976. "The urban informal sector", in *World Development,* Vol. 4, No. 8.

Mellor, J.W.; Johnston, B.F. 1984. "The world food equation: Inter-relations among development, employment and food consumption", *Journal of Economic Literature,* Vol. XXII, June.

Morawetz, D. 1974. "Employment implications of industrialisation in developing countries: A survey", in *Economic Journal,* Vol. 84, No. 335.

Mukhopadhyay, S.; Lim, C.P. (eds.). 1985. *Development and diversification of rural industries in Asia,* 2 vols. (Kuala Lumpur, APDC).

Nurkse, R. 1953. *Problems of capital formation in underdeveloped countries* (Oxford, Blackwell).

Prebisch, R. 1959. "Commercial policies in underdeveloped countries", in *American Economic Review,* Vol. 49, No. 2.

Pack, H. 1988. "Industrialisation and trade", in Chenery and Srinivasan (eds.), 1988.

Ranis, G.; Schultz, T.P. (eds.). 1988. *The state of development economics: Progress and perspectives* (Oxford, Blackwell).

Ray, S. 1987. *Returns to labour in Asia* (Geneva, ILO; mimeographed World Employment Programme working paper; restricted).

Rosensweig, M.R. 1988. "Labour markets in low-income countries: Distortions, mobility and migration", in Chenery and Srinivasan (eds.), 1988.

Saith, A. 1989a. *Development strategies and the rural poor* (Geneva, ILO; mimeographed World Employment Programme research working paper; restricted).

– – –. 1989b. *Rural non-farm economy: Processes and policies* (Geneva, ILO).

Salvatore, D. (ed.). 1988. *World population trends and their impact on economic development* (New York, Greenwood Press).

Scandizzo, P.L.; Bruce, C. 1980. "Methodologies for measuring agricultural price intervention effects", World Bank Staff Working Paper No. 396 (Washington, DC).

Schultz, T. 1964. *Transforming traditional agriculture* (New Haven, Connecticut, Yale University Press).

Sen, A. 1966. "Peasants and dualism with or without surplus labour", in *Journal of Political Economy*, Vol. 74, No. 5.

– – –. 1968. *Choice of techniques* (Oxford, Blackwell).

– – –. 1975. *Employment, technology and development* (Oxford, Clarendon Press).

Singer, H. 1950. "The distribution of gains between investing and borrowing countries", in *American Economic Review*, Vol. 40, No. 2.

– – –. 1987. "Terms of trade and economic development", in Eatwell et al. (eds.), 1987.

Squire, L. 1981. *Employment policy in developing countries: A survey of issues and evidence* (Oxford University Press for the World Bank).

Standing, G. 1978. *Labour force participation and development* (Geneva, ILO).

Steel, W.M.; Takagi, Y. 1983. "Small enterprise development and the employment-output trade-off", in *Oxford Economic Papers*, Vol. 35, No. 3.

Stolper, W.; Samuelson, P. 1941. "Protection and real wages", in *Review of Economic Studies*, 9 (1), Nov.

Timmer, C.P. 1988. "The agricultural transformation", in Chenery and Srinivasan (eds.), 1988.

Todaro, M.P. 1969. "A model of labour migration and urban unemployment in less developed countries", in *American Economic Review*, Vol. 59, No. 1.

– – –. 1976. *Internal migration in developing countries: A review of theory, evidence, methodology and research priorities* (Geneva, ILO).

Tokman, V.E. 1978. "An exploration into the nature of informal-formal sector relationships", in *World Development*, Sep.-Oct.

– – –. 1988. "Urban employment: Research and policy in Latin America", in *CEPAL Review*, Vol. 34, Apr.

Turnham, D. 1970. *The employment problem in less developed countries* (Paris, OECD).

Williamson, J.G. 1988. "Migration and urbanisation in the Third World", in Chenery and Srinivasan (eds.), 1988.

World Bank. 1983. *World Development Report* (New York, Oxford University Press).

– – –. 1984. *World Development Report* (New York, Oxford University Press).

– – –. 1986. *World Development Report* (New York, Oxford University Press).

– – –. 1987. *World Development Report* (New York, Oxford University Press).

Yap, L.Y.L. 1977. "The attraction of cities", in *Journal of Development Studies*, Vol. 4, No. 3.

Appendices

A. List of countries and regions

Region: *Sub-Saharan Africa*

Congo, Gabon, Nigeria, Togo, Niger, Zambia, Liberia, Guinea, Zaire, Mauritania, Ethiopia, Burkina Faso, Malawi, Mali, Madagascar, Mozambique, Burundi, United Republic of Tanzania, Somalia, Rwanda, Kenya, Sudan, Ghana, Senegal, Chad, Côte d'Ivoire, Benin, Gambia, Mauritius, Guinea Bissau, Sierra Leone, Cameroon, Zimbabwe, Central African Republic, Lesotho, Botswana, Swaziland, Angola, Uganda.

Region: *Asia-Pacific*

Myanmar, Sri Lanka, Nepal, Fiji, Indonesia, India, Bangladesh, China, Pakistan, Thailand, Philippines, Papua New Guinea, Republic of Korea, Malaysia, Afghanistan.

Region: *Latin America-Caribbean*

Trinidad-Tobago, Venezuela, Bolivia, Chile, Dominican Republic, Honduras, Nicaragua, El Salvador, Guatemala, Paraguay, Colombia, Costa Rica, Panama, Argentina, Guyana, Ecuador, Brazil, Peru, Mexico, Jamaica, Uruguay.

Region: *Near East-North Africa*

Algeria, Egypt, Islamic Republic of Iran, Iraq, Syrian Arab Republic, Libyan Arab Jamahiriya, Saudi Arabia, Yemen Arab Republic, Jordan, Morocco, Tunisia.

B. Statistical tables

General notes

In some tables, the growth rates are for the period 1970-79 for a few countries and for the period 1979-84 for a few others.

Except in the case of table X and for wage employment in table XII, the growth rates were estimated by fitting a semi-logarithmic trend line to annual data. When the coefficient was not significant even at 15 per cent, the growth rate was recorded as zero. In table X and for wage employment in table XII, growth rates were calculated from average values for 1969-71, 1978-80 and 1983-85.

The countries are listed in ascending order of per capita real GPD in 1970.

TABLE I : AVERAGE ANNUAL RATE OF GROWTH OF POPULATION AND LABOUR FORCE (PERCENTAGES)

COUNTRY	POPULATION		LABOUR FORCE	
	1969-79	1979-85	1969-79	1979-85
YEMEN ARAB REP.	2.03	2.68	0.69	2.53
ETHIOPIA	2.29	2.44	2.01	1.76
RWANDA	3.21	3.32	2.99	2.80
BURKINA FASO	1.90	2.35	1.69	1.92
MALI	2.10	2.74	1.68	2.42
MALAWI	2.72	3.06	2.16	2.57
AFGHANISTAN	1.86	0.59	1.66	0.74
LESOTHO	2.25	2.53	1.90	1.98
BURUNDI	1.67	2.71	1.30	1.92
CHAD	2.00	2.26	1.65	1.81
NEPAL	2.44	2.34	1.73	2.21
MYANMAR	2.22	1.95	2.21	1.95
INDONESIA	2.33	1.96	2.10	2.36
TANZANIA, UNITED REP. OF	3.30	3.51	2.78	2.77
CENT.AFR.REP	1.97	2.28	1.20	1.33
INDIA	2.19	1.95	1.69	1.97
BANGLADESH	2.79	2.75	1.99	2.69
BENIN	2.48	2.94	1.96	2.04
CHINA	1.94	1.25	2.42	2.44
GAMBIA	2.26	1.94	1.98	1.28
SOMALIA	4.13	3.07	3.63	2.17
TOGO	2.40	2.90	2.11	2.21
MADAGASCAR	2.55	2.79	2.12	1.95
BOTSWANA	3.78	3.81	2.97	3.38
UGANDA	2.88	3.31	2.59	2.72
KENYA	3.92	4.11	3.53	3.44
SUDAN	2.93	2.87	2.60	2.75
SIERRA LEONE	1.47	1.76	0.92	1.11
PAKISTAN	2.68	3.04	2.67	3.16
GUINEA	2.06	2.31	1.76	1.63
NIGER	2.42	2.80	1.82	2.21
SRI LANKA	1.67	1.80	2.13	1.76
CAMEROON	2.41	2.69	1.55	1.77
MAURITANIA	2.65	2.91	1.75	2.58
THAILAND	2.52	2.04	2.77	2.49
NIGERIA	3.40	3.35	3.02	2.66
ZAIRE	2.78	2.93	1.81	2.19
GUINEA BISSAU	4.07	2.12	3.63	1.37
PHILIPPINES	2.54	2.42	2.53	2.46
JORDAN	2.36	3.59	1.14	3.93
EGYPT	2.19	2.47	2.03	2.53
SENEGAL	3.45	2.65	3.21	2.00
MOZAMBIQUE	3.81	2.99	3.52	2.36

TABLE I (CONTD.): AVERAGE ANNUAL RATE OF GROWTH OF POPULATION AND LABOUR FORCE (PERCENTAGES)

COUNTRY	POPULATION		LABOUR FORCE	
	1969-79	1979-85	1969-79	1979-85
CONGO	2.39	2.58	2.06	1.86
BOLIVIA	2.50	2.68	2.05	2.60
GHANA	2.82	3.24	2.23	2.63
MOROCCO	2.40	2.46	3.19	3.29
PARAGUAY	3.21	3.03	3.43	3.11
MAURITIUS	1.00	1.93	2.25	3.30
COTE D'IVOIRE	3.90	3.67	2.60	2.63
PAPUA N.GUINEA	2.36	2.60	1.85	2.14
HONDURAS	3.29	3.41	3.02	3.74
KOREA, REP. OF	1.87	1.57	2.64	2.59
ECUADOR	2.97	2.87	2.62	3.00
TUNISIA	2.10	2.11	3.56	3.13
ZIMBABWE	3.26	3.49	2.80	2.72
SYRIAN ARAB REP	3.41	3.53	3.34	3.43
EL SALVADOR	2.94	2.92	2.96	2.88
ANGOLA	3.16	2.56	2.66	1.80
SWAZILAND	2.66	3.02	1.99	2.19
LIBERIA	3.08	3.17	2.58	2.19
COLOMBIA	2.18	2.15	2.51	2.77
DOMINICAN REP.	2.67	2.32	3.02	3.37
ALGERIA	3.04	3.04	3.05	3.51
GABON	1.11	1.54	0.82	0.61
GUATEMALA	2.77	2.81	2.18	2.71
GUYANA	1.95	1.97	3.59	3.10
NICARAGUA	3.04	3.28	2.91	3.60
IRAQ	3.45	3.60	3.87	3.71
MALAYSIA	2.41	2.43	3.65	2.98
IRAN, ISLAMIC REP.OF	3.12	2.89	3.13	3.23
FIJI	2.00	1.85	3.09	2.24
ZAMBIA	2.95	3.30	2.63	3.14
BRAZIL	2.38	2.24	3.39	2.42
PERU	2.73	2.61	3.23	2.92
COSTA RICA	2.69	2.67	3.76	3.13
PANAMA	2.53	2.17	2.47	2.86
SAUDI ARABIA	4.78	4.27	5.30	4.45
MEXICO	3.12	2.61	4.23	3.30
JAMAICA	1.57	1.42	2.67	2.95
URUGUAY	0.30	0.70	0.14	0.64
CHILE	1.68	1.56	2.36	2.54
TRINIDAD TOBAGO	1.30	1.59	2.29	2.49
ARGENTINA	1.64	1.59	1.04	1.07
VENEZUELA	3.49	2.90	4.68	3.56
LIBYAN ARAB JAMAHIRIYA	4.05	3.87	3.75	3.62

SOURCE : FAO AND ILO.

TABLE II : AVERAGE ANNUAL RATE OF GROWTH OF AGRICULTURAL AND NON AGRICULTURAL LABOUR FORCE (PERCENTAGES)

COUNTRY	AGRICULTURAL		NON_AGRICULTURAL	
	1969-79	1979-85	1969-79	1979-85
YEMEN ARAB REP.	-0.33	1.61	3.53	4.45
ETHIOPIA	1.40	1.12	4.96	4.14
RWANDA	2.90	2.65	4.32	4.62
BURKINA FASO	1.51	1.67	2.98	3.48
MALI	1.27	1.90	4.58	5.23
MALAWI	1.36	1.64	7.85	6.77
AFGHANISTAN	0.87	-0.29	3.08	2.26
LESOTHO	1.48	1.31	5.13	5.75
BURUNDI	1.22	1.77	2.43	3.71
CHAD	0.88	0.85	7.05	6.12
NEPAL	1.66	2.08	2.74	3.78
MYANMAR	1.11	0.78	3.65	3.21
INDONESIA	0.65	0.78	4.54	4.32
TANZANIA, UNITED REP. OF	2.26	2.24	6.84	5.72
CENT.AFR.REP.	-0.14	-0.01	6.23	4.57
INDIA	1.40	1.52	2.38	2.96
BANGLADESH	1.16	1.86	5.08	5.02
BENIN	0.59	0.76	6.45	4.82
CHINA	1.88	1.58	4.20	4.75
GAMBIA	1.68	0.94	3.75	2.97
SOMALIA	3.14	1.50	5.35	4.10
TOGO	1.61	1.74	3.61	3.42
MADAGASCAR	1.78	1.45	3.76	3.97
BOTSWANA	1.08	2.26	10.45	5.89
UGANDA	2.20	2.19	5.41	5.74
KENYA	3.08	2.96	5.78	5.40
SUDAN	1.80	1.29	4.98	5.99
SIERRA LEONE	0.11	0.08	3.14	3.31
PAKISTAN	1.95	2.25	3.63	4.20
GUINEA	1.23	0.89	4.43	4.42
NIGER	1.49	1.83	6.32	5.83
SRI LANKA	1.78	1.44	2.54	2.12
CAMEROON	-0.16	0.45	7.65	4.61
MAURITANIA	-0.19	1.77	8.99	4.36
THAILAND	1.63	1.53	6.42	4.70
NIGERIA	2.62	2.18	3.95	3.66
ZAIRE	0.83	1.36	4.95	4.18
GUINEA BISSAU	3.41	0.93	4.72	3.31
PHILIPPINES	1.94	1.50	3.20	3.45
JORDAN	-8.15	-2.08	3.42	4.56
EGYPT	0.76	1.34	3.28	3.50
SENEGAL	2.96	1.74	4.34	3.04
MOZAMBIQUE	3.29	2.03	4.87	4.05

TABLE II (CONTD.): AVERAGE ANNUAL RATE OF GROWTH OF AGRICULTURAL AND NON AGRICULTURAL LABOUR FORCE (PERCENTAGES)

COUNTRY	AGRICULTURAL		NON_AGRICULTURAL	
	1969-79	1979-85	1969-79	1979-85
CONGO	1.65	1.40	2.78	2.60
BOLIVIA	0.92	1.48	3.17	3.52
GHANA	1.76	1.62	2.86	3.84
MOROCCO	0.93	1.18	5.70	4.94
PARAGUAY	2.64	2.54	4.26	3.65
MAURITIUS	0.31	1.28	3.14	4.04
COTE D'IVOIRE	1.05	1.12	6.60	5.26
PAPUA N.GUINEA	0.87	0.97	6.08	5.57
HONDURAS	2.32	2.85	4.23	5.05
KOREA, REP. OF	-0.27	-1.12	4.92	4.48
ECUADOR	0.00	0.61	4.82	4.39
TUNISIA	1.56	0.00	4.84	4.77
ZIMBABWE	2.21	2.09	4.62	4.34
SYRIAN ARAB REP	-0.84	0.00	6.41	4.92
EL SALVADOR	0.45	1.35	5.53	3.99
ANGOLA	2.14	1.27	4.30	3.25
SWAZILAND	1.15	1.20	5.00	4.81
LIBERIA	2.17	1.66	3.91	3.67
COLOMBIA	1.09	0.75	3.36	3.76
DOMINICAN REP.	1.23	1.05	4.88	5.16
ALGERIA	-1.02	0.40	5.84	4.82
GABON	0.28	-0.31	2.72	3.23
GUATEMALA	1.42	1.72	3.30	3.95
GUYANA	1.83	1.32	4.33	3.72
NICARAGUAY	1.87	1.84	3.94	5.04
IRAQ	-0.68	-0.37	6.79	5.28
MALAYSIA	1.16	0.47	6.01	4.64
IRAN, ISLAMIC REP.OF	1.24	0.40	4.42	4.69
FIJI	1.94	0.61	4.23	3.55
ZAMBIA	2.15	2.57	4.06	4.63
BRAZIL	-0.11	-0.17	5.63	3.52
PERU	1.63	1.50	4.50	3.83
COSTA RICA	0.64	0.50	5.63	4.23
PANAMA	-0.14	0.51	4.03	3.89
SAUDI ARABIA	2.60	2.32	8.97	6.31
MEXICO	2.35	1.37	5.55	4.35
JAMAICA	1.99	1.65	2.99	3.52
URUGUAY	-1.50	-0.85	0.48	0.91
CHILE	-1.00	-0.26	3.22	3.06
TRINIDAD TOBAGO	-3.35	-0.99	3.28	2.86
ARGENTINA	-1.02	-1.18	1.40	1.39
VENEZUELA	0.06	-0.34	5.95	4.25
LIBYAN ARAB JAMAHARIYA	-0.93	-0.96	5.26	4.53

SOURCE: FAO AND ILO.

TABLE III : PERCENTAGE SHARAB REP.E OF AGRICULTURE IN LABOUR FORCE : ABSOLUTE VALUES AND AVERAGE ANNUAL RATE OF CHANGE (PERCENTAGES)

COUNTRY	PERCENTAGE SHARAB REP.E			RATE OF CHANGE	
	1969	1979	1985	1969-79	1979-85
YEMEN ARAB REP.	76.9	69.7	65.8	-1.03	-0.92
ETHIOPIA	85.2	80.3	77.3	-0.61	-0.64
RWANDA	93.8	92.9	92.1	-0.10	-0.15
BURKINA FASO	88.5	86.8	85.6	-0.18	-0.25
MALI	89.4	85.9	83.3	-0.41	-0.51
MALAWI	90.8	84.1	79.5	-0.80	-0.93
AFGHANISTAN	66.6	61.5	57.9	-0.80	-1.03
LESOTHO	90.3	86.6	83.2	-0.42	-0.67
BURUNDI	93.6	92.8	92.0	-0.08	-0.15
CHAD	90.5	84.0	79.3	-0.77	-0.96
NEPAL	93.7	93.0	92.4	-0.07	-0.12
MYANMAR	60.0	53.6	50.0	-1.10	-1.17
INDONESIA	67.1	58.1	52.8	-1.45	-1.58
TANZANIA, UNITED REP. OF	90.6	86.1	83.4	-0.53	-0.53
CENT.AFR.REP.	84.0	73.5	67.7	-1.34	-1.35
INDIA	72.0	69.9	68.1	-0.28	-0.45
BANGLADESH	81.9	75.5	71.8	-0.83	-0.83
BENIN	81.3	71.2	65.9	-1.37	-1.28
CHINA	78.8	74.6	71.0	-0.54	-0.86
GAMBIA	86.7	84.2	82.4	-0.30	-0.34
SOMALIA	79.7	75.9	73.0	-0.49	-0.66
TOGO	77.0	73.4	71.5	-0.50	-0.46
MADAGASCAR	84.0	81.2	78.8	-0.34	-0.50
BOTSWANA	86.3	71.7	66.8	-1.90	-1.12
UGANDA	89.6	86.2	83.5	-0.39	-0.54
KENYA	85.1	81.4	79.1	-0.45	-0.48
SUDAN	77.9	71.7	65.9	-0.80	-1.46
SIERRA LEONE	76.1	70.2	66.0	-0.81	-1.03
PAKISTAN	59.0	55.0	52.1	-0.72	-0.90
GUINEA	85.5	81.1	77.6	-0.53	-0.75
NIGER	94.5	91.4	89.4	-0.34	-0.38
SRI LANKA	55.4	53.5	52.5	-0.34	-0.32
CAMEROON	84.0	71.2	65.6	-1.71	-1.32
MAURITANIA	85.6	70.8	67.1	-1.94	-0.81
THAILAND	80.1	71.8	67.7	-1.14	-0.96
NIGERIA	71.2	68.4	66.5	-0.40	-0.48
ZAIRE	79.7	72.3	68.7	-0.98	-0.83
GUINEA BISSAU	84.4	82.6	80.5	-0.22	-0.44
PHILIPPINES	55.4	52.1	49.2	-0.58	-0.96
JORDAN	29.4	11.5	7.9	-9.30	-6.01
EGYPT	52.6	46.3	43.1	-1.27	-1.19
SENEGAL	82.7	80.8	79.6	-0.25	-0.26
MOZAMBIQUE	86.6	84.7	83.1	-0.23	-0.32

TABLE III (CONTD.): PERCENTAGE SHARAB REP.E OF AGRICULTURE IN LABOUR FORCE : ABSOLUTE VALUES AND AVERAGE ANNUAL RATE OF CHANGE (PERCENTAGES)

COUNTRY	PERCENTAGE SHARAB REP.E			RATE OF CHANGE	
	1969	1979	1985	1969-79	1979-85
CONGO	65.1	62.6	61.0	-0.41	-0.46
BOLIVIA	52.5	46.9	43.9	-1.13	-1.11
GHANA	58.9	56.1	52.9	-0.47	-1.01
MOROCCO	58.2	46.7	41.1	-2.26	-2.11
PARAGUAY	53.0	49.0	47.2	-0.79	-0.58
MAURITIUS	34.7	28.6	25.3	-1.34	-2.02
COTE D'IVOIRE	77.3	66.3	60.5	-1.55	-1.51
PAPUA.N.GUINEA	84.8	77.0	71.9	-0.98	-1.17
HONDURAS	65.4	60.9	57.8	-0.70	-0.89
KOREA, REP. OF	50.2	37.6	30.1	-2.91	-3.72
ECUADOR	51.4	39.7	34.3	-2.63	-2.38
TUNISIA	43.4	35.4	29.1	-2.00	-3.43
ZIMBABWE	77.7	73.3	70.5	-0.59	-0.63
SYRIAN ARAB REP	50.5	33.8	27.4	-4.18	-3.41
EL SALVADOR	56.6	44.3	40.3	-2.51	-1.53
ANGOLA	78.0	74.2	71.8	-0.51	-0.53
SWAZILAND	81.5	74.5	70.3	-0.85	-1.00
LIBERIA	77.7	74.5	72.3	-0.42	-0.54
COLOMBIA	40.3	34.7	30.8	-1.42	-2.02
DOMINICAN REP.	55.6	46.5	40.6	-1.79	-2.32
ALGERIA	49.1	32.6	26.9	-4.07	-3.11
GABON	80.2	75.7	71.8	-0.54	-0.93
GUATEMALA	61.9	57.3	54.0	-0.76	-0.99
GUYANA	32.5	27.3	24.5	-1.76	-1.78
NICARAGUA	52.4	47.0	42.4	-1.04	-1.77
IRAQ	47.6	31.0	24.5	-4.55	-4.08
MALAYSIA	54.7	42.8	36.7	-2.48	-2.52
IRAN, ISLAMIC REP.OF	44.7	37.0	31.3	-1.88	-2.82
FIJI	52.5	46.6	42.4	-1.16	-1.63
ZAMBIA	77.0	73.4	71.0	-0.47	-0.57
BRAZIL	45.6	32.4	27.6	-3.50	-2.58
PERU	47.6	40.7	37.3	-1.60	-1.42
COSTA RICA	43.3	31.9	27.1	-3.12	-2.63
PANAMA	42.3	32.6	28.3	-2.61	-2.36
SAUDI ARABIA	64.9	49.9	43.8	-2.70	-2.12
MEXICO	45.2	37.3	33.2	-1.88	-1.93
JAMAICA	33.9	31.4	29.1	-0.68	-1.30
URUGUAY	18.8	16.0	14.6	-1.64	-1.49
CHILE	23.8	17.1	14.4	-3.37	-2.81
TRINIDAD TOBAGO	18.9	10.9	8.7	-5.64	-3.48
ARGENTINA	16.4	13.3	11.6	-2.06	-2.25
VENEZUELA	26.6	16.9	13.3	-4.62	-3.91
LIBYAN ARAB JAMAHIRIYA	30.8	19.0	14.4	-4.68	-4.58

SOURCE: FAO AND ILO.

TABLE IV : AVERAGE ANNUAL GROWTH RATE OF TOTAL GDP AND AGRICULTURAL GDP, 1980 PRICES (PERCENTAGES)

COUNTRY	TOTAL GDP		AGRICULTURAL GDP	
	1969-79	1979-85	1969-79	1979-85
YEMEN ARAB REP.	8.25	6.53	3.76	0.00
ETHIOPIA	1.69	0.00	0.52	0.00
RWANDA	3.46	2.71	6.82	0.00
BURKINA FASO	3.58	2.18	1.62	1.81
MALI	5.08	-0.85	4.13	-3.62
MALAWI	6.06	1.59	4.78	0.00
AFGHANISTAN	.	.	.	.
LESOTHO	7.61	0.00	0.00	0.00
BURUNDI	2.34	2.70	1.54	0.00
CHAD	.	.	.	.
NEPAL	.	.	0.76	4.05
MYANMAR	.	.	3.71	5.69
INDONESIA	7.68	4.28	3.81	3.35
TANZANIA, UNITED REP.OF	2.97	0.90	0.86	0.89
CENT.AFR.REP.	2.53	0.00	2.17	1.61
INDIA	3.63	5.28	1.86	3.60
BANGLADESH	3.23	3.35	1.91	2.49
BENIN	2.06	4.13	1.59	0.00
CHINA	.	.	2.90	7.24
GAMBIA	4.47	0.00	3.76	0.00
SOMALIA	.	.	.	.
TOGO	3.49	0.00	1.28	0.00
MADAGASCAR	.	.	0.00	1.59
BOTSWANA	12.71	11.21	9.21	-10.63
UGANDA	0.00	0.00	0.00	0.00
KENYA	6.01	3.36	4.74	2.85
SUDAN	6.06	0.00	4.29	0.00
SIERRA LEONE	1.52	2.33	2.89	0.77
PAKISTAN	4.61	6.38	2.03	3.00
GUINEA	.	.	-4.89	.
NIGER	.	.	.	1.74
SRI LANKA	3.59	.	2.57	3.95
CAMEROON	5.74	4.92	3.49	3.64
MAURITANIA	1.03	0.87	0.00	0.00
THAILAND	7.16	4.81	4.89	3.30
NIGERIA	5.72	-3.34	0.00	0.00
ZAIRE	0.00	1.62	1.34	2.48
GUINEA BISSAU	3.03	.	1.12	1.44
PHILIPPINES	5.94	0.00	4.82	2.16
JORDAN	7.94	6.98	-4.09	6.08
EGYPT	8.02	6.18	2.73	2.08
SENEGAL	2.41	.	2.27	0.00

TABLE IV (CONTD.): AVERAGE ANNUAL GROWTH RATE OF TOTAL GDP AND AGRICULTURAL GDP, 1980 PRICES (PERCENTAGES)

COUNTRY	TOTAL GDP		AGRICULTURAL GDP	
	1969-79	1979-85	1969-79	1979-85
MOZAMBIQUE	.	.	.	.
CONGO	.	.	2.16	0.00
BOLIVIA	4.58	.	4.11	0.00
GHANA	0.00	.	0.00	0.00
MOROCCO	.	.	0.00	0.00
PARAGUAY	7.90	3.22	5.95	2.90
MAURITIUS	8.23	2.58	0.00	0.00
COTE D'IVOIRE	5.26	0.00	2.14	0.00
PAPUA.N.GUINEA	.	.	2.60	2.64
HONDURAS	4.13	0.81	1.45	2.25
KOREA, REP.OF	9.19	6.14	4.02	3.81
ECUADOR	10.18	1.65	2.86	0.00
TUNISIA	6.04	4.49	4.38	4.21
ZIMBABWE	1.35	3.71	0.00	0.00
SYRIAN ARAB REP	.	.	7.38	0.00
EL SALVADOR	4.51	.	3.63	-3.51
ANGOLA	.	.	.	.
SWAZILAND	4.36	2.26	.	.
LIBERIA	2.02	-1.69	4.81	0.00
COLOMBIA	4.96	2.37	4.52	1.77
DOMINICAN REP.	6.81	3.18	3.29	2.64
ALGERIA	7.29	4.22	6.86	0.00
GABON	.	.	.	.
GUATEMALA	.	.	.	.
GUYANA	2.03	-3.86	1.50	0.00
NICARAGUA	3.99	.	4.72	0.00
IRAQ	.	.	.	.
MALAYSIA	6.56	5.67	5.15	2.95
IRAN, ISLAMIC REP.OF	.	.	.	.
FIJI	4.91	0.00	3.15	0.00
ZAMBIA	1.61	0.00	2.61	1.60
BRAZIL	8.15	1.82	3.89	3.46
PERU	3.68	0.00	0.00	0.00
COSTA RICA	.	0.00	2.66	2.04
PANAMA	.	2.48	2.27	2.13
SAUDI ARABIA	10.68	.	4.91	8.39
MEXICO	5.85	.	3.06	2.93
JAMAICA	-1.00	0.00	0.00	0.00
URUGUAY	2.80	-2.84	0.00	0.00
CHILE	0.00	.	2.86	2.15
TRINIDAD TOBAGO	4.89	-3.36	-0.94	3.28
ARGENTINA	2.32	-1.68	2.92	2.00
VENEZUELA	.	.	3.38	1.44
LIBYAN ARAB JAMAHIRIYA	2.14	-6.36	11.25	7.65

SOURCE: WORLD BANK.

TABLE V : AVERAGE ANNUAL GROWTH RATE OF NON-AGRICULTURAL GDP AND MANUFACTURING GDP, 1980 PRICES (PERCENTAGES)

COUNTRY	NON-AGRICULTURAL GDP		MANUFACTURING GDP	
	1969-79	1979-85	1969-79	1979-85
YEMEN ARAB REP	11.52	9.07	11.37	15.52
ETHIOPIA	3.09	3.89	1.92	4.24
RWANDA	0.00	5.73	3.81	6.39
BURKINA FASO	5.41	2.45	.	.
MALI	6.97	3.12	.	.
MALAWI	6.89	1.89	.	.
AFGHANISTAN	.	.	.	.
LESOTHO	12.25	0.00	16.54	14.11
BURUNDI	3.89	4.52	3.29	6.67
CHAD	.	.	.	.
NEPAL	.	.	.	.
BURMA	.	.	3.86	5.76
INDONESIA	9.30	4.55	12.60	8.39
TANZANIA, UNITED REP.OF	5.03	0.90	4.16	-5.29
CENT.AFR.REP.	2.78	0.00	4.83	-2.29
INDIA	4.84	6.20	4.74	5.06
BANGLADESH	5.05	4.27	11.64	2.13
BENIN	2.51	6.49	0.00	6.45
CHINA	.	.	6.58	10.20
GAMBIA	4.77	3.44	.	.
SOMALIA	.	.	.	.
TOGO	4.58	0.00	0.00	0.00
MADAGASCAR	.	.	.	.
BOTSWANA	13.59	13.41	19.11	3.99
UGANDA	-2.62	2.49	-8.73	2.45
KENYA	6.73	3.61	9.90	3.82
SUDAN	7.31	1.66	3.79	0.00
SIERRA LEONE	0.86	3.09	4.62	0.00
PAKISTAN	5.91	7.64	4.77	9.50
GUINEA	.	.	.	.
NIGER	.	.	.	.
SRI LANKA	4.02	5.28	1.66	4.78
CAMEROON	6.99	9.29	6.39	17.14
MAURITANIA	2.41	0.00	.	.
THAILAND	8.34	5.39	10.31	5.23
NIGERIA	7.94	-4.66	13.16	5.12
ZAIRE	0.00	1.24	0.00	-1.60
GUINEA BISSAU	7.07	.	.	.
PHILIPPINES	6.37	0.00	6.92	0.00
JORDAN	9.57	7.06	16.20	7.09
EGYPT	10.07	7.15	.	.
SENEGAL	2.45	.	2.79	4.32
MOZAMBIQUE	.	.	.	.

TABLE V (CONTD.) : AVERAGE ANNUAL GROWTH RATE OF NON-AGRICULTURAL GDP AND MANUFACTURING GDP, 1980 PRICES (PERCENTAGES)

COUNTRY	NON-AGRICULTURAL GDP		MANUFACTURING GDP	
	1969-79	1979-85	1969-79	1979-85
CONGO	.	.	0.00	8.47
BOLIVIA	4.69	.	6.37	-9.25
GHANA	2.67	.	0.00	-6.92
MOROCCO	.	.	5.77	1.19
PARAGUAY	8.99	3.36	7.14	0.00
MAURITIUS	11.56	2.85	7.73	4.72
COTE D'IVOIRE	7.58	0.00	7.03	7.31
PAPUA N.GUINEA	.	.	.	.
HONDURAS	5.35	0.00	5.27	-1.35
KOREA, REP.OF	10.97	6.65	16.69	7.39
ECUADOR	11.83	1.79	10.33	0.00
TUNISIA	6.42	4.54	9.67	7.57
ZIMBABWE	1.48	3.77	2.41	2.70
SYRIAN ARAB RP	.	.	.	.
EL SALVADOR	4.87	.	5.14	-3.94
ANGOLA	.	.	.	.
SWAZILAND	.	.	.	.
LIBERIA	0.77	-3.13	8.43	-7.38
COLOMBIA	5.10	2.54	4.43	2.42
DOMINICAN REP.	7.85	3.05	6.64	1.46
ALGERIA	7.33	4.36	8.47	8.57
GABON	.	.	.	.
GUATEMALA	.	.	.	.
GUYANA	3.20	-4.15	5.54	-7.08
NICARAGUA	3.61	.	3.80	2.47
IRAQ	.	.	.	.
MALAYSIA	7.15	6.62	11.24	6.21
IRAN, ISLAMIC REP.OF	.	.	.	.
FIJI	5.50	1.09	4.82	0.00
ZAMBIA	1.42	0.00	2.69	0.00
BRAZIL	8.79	1.61	8.37	0.00
PERU	4.26	0.00	3.07	-2.67
COSTA RICA	.	0.00	.	.
PANAMA	.	2.47	2.42	0.00
SAUDI ARABIA	10.77	.	5.79	7.54
MEXICO	6.19	.	6.64	0.00
JAMAICA	-1.15	0.00	0.00	0.00
URUGUAY	3.16	-3.29	.	.
CHILE	0.00	.	0.00	0.00
TRINIDAD TOBAGO	5.10	-3.84	1.47	-14.00
ARGENTINA	2.26	-2.08	1.60	-2.85
VENEZUELA	.	.	5.75	1.33
LIBYAN ARAB JAMAHIRIYA	0.00	-6.66	15.15	7.41

SOURCE: WORLD BANK.

TABLE VI : FOOD CONSUMPTION (CALORIES) PER CAPITA PER DAY ABSOLUTE VALUES AND AVERAGE ANNUAL GROWTH RATES (PERCENTAGES)

COUNTRY	FOOD CONSUMPTION IN CALORIES			
	VALUE		GROWTH RATE	
	1969	1985	1969-79	1979-85
YEMEN ARAB REP.	1991.0	2109.0	-1.04	1.36
ETHIOPIA	1781.4	1521.1	-0.66	-1.14
RWANDA	1756.3	2021.8	1.70	0.00
BURKINA FASO	1743.4	1991.0	0.69	1.36
MALI	1794.2	1785.3	0.00	0.00
MALAWI	2045.2	2520.1	1.42	0.89
AFGHANISTAN	2197.3	2223.0	0.00	0.41
LESOTHO	2008.4	2310.0	0.00	2.27
BURUNDI	2318.0	2395.2	0.00	0.00
CHAD	2246.8	1826.1	-0.91	0.00
NEPAL	1848.7	1924.4	0.72	0.00
BURMA	1787.9	2146.6	1.60	0.79
INDONESIA	1668.6	2266.5	1.78	1.53
TANZANIA, UNITED REP.OF	1912.7	2407.4	0.00	4.67
CENT.AFR.REP.	2018.9	2155.6	0.54	0.00
INDIA	2087.7	2034.0	0.00	0.00
BANGLADESH	1007.4	1825.8	0.43	-0.76
BENIN	2071.3	2114.0	0.53	0.00
CHINA	1726.1	2093.0	1.74	0.52
GAMBIA	2141.3	2193.0	0.57	0.00
SOMALIA	1951.4	2015.4	1.19	0.00
TOGO	2108.7	2014.7	0.00	-1.20
MADAGASCAR	2380.3	2473.2	0.43	0.00
BOTSWANA	2056.2	2152.0	0.66	0.00
UGANDA	2360.6	2288.2	0.00	0.00
KENYA	2247.8	2262.7	0.00	0.00
SUDAN	1791.8	2246.1	2.09	0.00
SIERRA LEONE	1580.3	1955.5	2.43	0.00
PAKISTAN	1698.2	2146.0	2.17	1.08
GUINEA	1788.9	1904.7	0.65	0.00
NIGER	2081.2	2196.1	0.00	0.00
SRI LANKA	2111.2	2343.6	1.01	0.00
CAMEROON	2032.7	2246.9	0.85	0.00
MAURITANIA	2023.6	1961.1	0.00	0.00
THAILAND	2071.5	2374.5	1.31	0.98
NIGERIA	2213.0	2169.7	-0.33	0.00
ZAIRE	2150.2	2242.5	0.57	0.00
GUINEA BISSAU	1730.9	1764.1	1.12	-1.02
PHILIPPINES	1784.9	2237.5	1.43	1.55
JORDAN	2161.2	1795.9	-3.00	0.00
EGYPT	2309.3	2768.3	0.81	1.78
SENEGAL	2268.3	2354.3	0.00	0.00
MOZAMBIQUE	2008.9	1898.5	0.34	-0.80

TABLE VI(CONTD.): FOOD CONSUMPTION (CALORIES) PER CAPITA PER DAY ABSOLUTE VALUES AND AVERAGE ANNUAL GROWTH RATES (PERCENTAGES)

COUNTRY	FOOD CONSUMPTION IN CALORIES			
	VALUE		GROWTH RATE	
	1969	1985	1969-79	1979-85
CONGO	2194.9	2336.4	0.00	0.64
BOLIVIA	1764.0	2041.2	1.26	0.59
GHANA	2117.8	1965.7	1.09	0.00
MOROCCO	2157.5	2718.7	0.98	1.27
PARAGUAY	2427.7	2771.0	1.48	0.00
MAURITIUS	2386.3	2682.1	-0.50	2.22
COTE D'IVOIRE	1983.3	2334.8	1.18	0.00
PAPUA N.GUINEA	1831.3	2071.7	1.41	0.00
HONDURAS	1939.5	2137.7	1.44	0.00
KOREA, REP.OF	2026.5	2765.7	2.57	1.23
ECUADOR	1792.7	2036.5	0.67	0.96
TUNISIA	2124.3	2632.0	0.91	1.44
ZIMBABWE	2114.3	2173.3	0.00	0.00
SYRIAN ARAB REP	2309.0	2529.3	0.00	0.51
EL SALVADOR	1725.6	2115.6	0.66	2.36
ANGOLA	1780.8	2150.6	1.35	0.00
SWAZILAND	2141.9	2422.4	0.65	1.69
LIBERIA	2066.1	2386.9	0.68	1.03
COLOMBIA	2263.4	2459.6	-0.47	1.83
DOMINICAN REP.	1746.4	2272.9	1.45	1.23
ALGERIA	1792.7	2338.6	0.43	2.87
GABON	1956.7	2057.9	-0.35	2.10
GUATEMALA	1903.7	2174.8	0.92	0.72
GUYANA	2268.1	2371.4	0.00	0.00
NICARAGUA	2239.4	2436.8	0.86	0.00
IRAQ	2053.0	2424.9	1.07	1.57
MALAYSIA	2284.0	2551.0	0.73	0.86
IRAN, ISLAMIC REP.OF	1970.4	2894.4	1.49	4.42
FIJI	2481.3	2575.5	0.50	-1.23
ZAMBIA	2121.8	2352.0	0.64	0.94
BRAZIL	2281.3	2516.9	0.83	0.32
PERU	2271.6	2208.3	0.00	-0.48
COSTA RICA	2168.0	2604.4	1.03	1.06
PANAMA	2264.0	2364.7	0.60	0.00
SAUDI ARABIA	1811.9	2194.0	0.41	2.11
MEXICO	2558.1	2870.7	0.68	1.03
JAMAICA	1986.9	2664.2	2.59	0.00
URUGUAY	2203.7	2761.8	0.87	0.00
CHILE	2497.3	2526.9	0.66	-1.23
TRINIDAD TOBAGO	2351.1	2657.1	0.77	0.45
ARGENTINA	3183.9	3216.7	0.70	0.00
VENEZUELA	2237.1	2612.2	0.90	1.65
LIBYAN ARAB JAMAHIRIYA	1691.0	3451.1	4.29	7.15

SOURCE: FAO.

TABLE VII : TECHNICAL CONDITIONS OF PRODUCTION IN AGRICULTURE - I

COUNTRY	ARABLE LAND PER AGRICULTURAL WORKER			% OF ARABLE LAND IRRIGATED		
	1969 (HA)	GROWTH RATE (%) 1969-79	GROWTH RATE (%) 1979-85	1969	GROWTH RATE (%) 1969-79	GROWTH RATE (%) 1979-85
YEMEN ARAB REP	1.27	0.71	-1.64	15.6	1.05	0.18
ETHIOPIA	1.10	-0.99	-1.03	0.5	2.93	0.00
RWANDA	0.34	0.75	-1.69	1.0	3.83	1.48
BURKINA FASO	0.89	0.00	-1.14	0.1	16.12	0.00
MALI	1.00	0.94	-1.70	7.6	4.01	6.34
MALAWI	1.07	-0.34	-1.17	0.2	16.18	2.10
AFGHANISTAN	2.97	-0.82	0.32	29.0	1.27	0.09
LESOTHO	0.75	-3.79	-1.32	.	.	.
BURUNDI	0.67	-0.45	-1.68	0.4	0.00	0.00
CHAD	2.36	0.00	-0.85	0.2	0.00	0.00
NEPAL	0.39	0.64	-2.02	6.2	13.82	5.42
MYANMAR	1.46	-1.63	-0.64	7.9	2.61	0.00
INDONESIA	0.60	0.30	0.52	23.7	1.41	2.60
TANZANIA, UNITED REP.OF	0.75	-1.43	-2.11	0.8	9.89	4.11
CENT.AFR.REP.	2.07	0.74	0.46	.	.	.
INDIA	1.04	-1.14	-1.37	17.6	2.19	0.86
BANGLADESH	0.55	-1.10	-1.91	16.1	2.31	5.45
BENIN	1.35	0.86	-0.40	0.6	5.76	2.72
CHINA	0.31	-2.13	-1.33	36.0	2.39	-0.38
GAMBIA	0.64	0.34	0.44	15.4	2.17	0.00
SOMALIA	1.05	-2.67	-1.58	11.2	1.00	4.60
TOGO	2.03	-1.34	-1.60	0.3	0.00	-2.36
MADAGASCAR	0.86	1.00	-1.21	14.0	6.34	4.78
BOTSWANA	5.64	0.83	-2.27	0.4	7.73	3.60
UGANDA	1.21	-0.76	0.84	0.2	5.16	-7.24
KENYA	0.50	-1.97	-2.33	1.3	3.20	-1.02
SUDAN	3.25	-1.09	-1.22	14.4	-2.14	1.20
SIERRA LEONE	1.69	1.65	0.11	0.4	8.60	0.00
PAKISTAN	1.72	-1.39	-2.06	66.4	0.67	1.05
GUINEA	0.85	-1.25	-0.83	3.1	3.38	0.49
NIGER	1.21	0.00	0.00	0.1	13.09	6.16
SRI LANKA	0.84	-0.79	-0.83	20.4	1.26	0.00
CAMEROON	2.33	1.67	-0.33	0.1	0.00	0.00
MAURITANIA	0.78	-4.17	-1.93	1.4	12.22	0.00
THAILAND	0.97	1.55	0.00	13.6	0.62	2.57
NIGERIA	1.82	-2.46	-1.72	0.1	36.59	11.12
ZAIRE	0.86	-0.20	-0.60	.	0.00	0.00
GUINEA BISSAU	1.18	-2.93	-0.62	.	.	.
PHILIPPINES	0.95	-1.02	-1.07	11.3	3.01	3.22
JORDAN	2.15	9.06	3.47	9.5	0.00	0.00
EGYPT	0.60	-2.16	-1.08	100.0	0.00	-0.02
SENEGAL	3.06	-1.96	-1.72	2.3	4.17	0.00
MOZAMBIQUE	0.73	-2.92	-2.16	0.8	8.62	5.83

TABLE VII (CONTD.) : TECHNICAL CONDITIONS OF PRODUCTION IN AGRICULTURE - I

COUNTRY	ARABLE LAND PER AGRICULTURAL WORKER			% OF ARABLE LAND IRRIGATED		
	1969 (HA)	GROWTH RATE (%)		1969	GROWTH RATE (%)	
		1969-79	1979-85		1969-79	1979-85
CONGO	1.87	-0.93	-1.19	0.2	8.69	8.69
BOLIVIA	2.88	4.14	-1.43	3.8	0.00	2.81
GHANA	1.32	-1.18	-1.28	0.2	0.00	-7.24
MOROCCO	3.21	-0.30	-0.24	4.4	3.66	-0.51
PARAGUAY	2.39	3.68	2.52	4.4	-3.16	-2.91
MAURITIUS	1.20	0.00	-1.23	14.4	0.00	1.25
COTE D'IVOIRE	1.33	3.11	-0.19	0.6	8.52	1.66
PAPUA N.GUINEA	0.33	-0.39	0.00	.	.	.
HONDURAS	3.03	-0.76	-2.64	4.3	0.52	0.76
KOREA, REP.OF	0.42	0.00	0.77	41.5	1.95	1.47
ECUADOR	2.71	-0.20	0.00	18.2	1.54	0.46
TUNISIA	7.75	-0.30	0.00	1.9	4.18	7.42
ZIMBABWE	1.37	-1.40	-0.89	2.0	8.39	7.08
SYRIAN ARAB RP	7.69	0.00	0.00	9.3	0.00	3.30
EL SALVADOR	0.97	0.77	-1.25	3.2	12.30	0.00
ANGOLA	1.73	-2.02	-1.26	.	.	.
SWAZILAND	0.96	0.00	0.00	29.6	0.00	0.00
LIBERIA	0.87	-2.10	-1.49	0.5	2.56	6.99
COLOMBIA	2.07	0.00	-0.51	4.9	0.99	0.83
DOMINICAN REP.	1.80	1.06	0.00	11.1	0.46	2.59
ALGERIA	4.71	0.32	0.00	3.5	-0.56	4.30
GABON	0.72	5.33	0.41	.	.	.
GUATEMALA	1.62	-0.21	-0.65	3.5	0.00	1.51
GUYANA	5.73	0.94	-1.31	31.3	-2.10	0.64
NICARAGUA	3.83	-1.56	-1.44	2.9	7.90	0.61
IRAQ	4.51	1.58	0.37	29.1	0.76	0.13
MALAYSIA	1.98	-0.15	-0.18	5.9	1.61	0.71
IRAN ISLAMIC REP.OF	4.46	-1.65	0.00	32.5	1.14	2.23
FIJI	2.91	-1.47	-0.30	0.4	0.00	0.00
ZAMBIA	4.49	-1.94	-2.24	0.1	11.50	0.00
BRAZIL	3.83	2.70	1.68	1.4	5.83	3.28
PERU	1.51	0.75	-0.37	39.4	-1.78	-0.34
COSTA RICA	2.21	0.98	0.00	5.3	7.06	11.16
PANAMA	2.59	0.28	0.00	3.7	3.27	0.83
SAUDI ARABIA	0.84	0.00	-1.40	43.2	-2.54	0.33
MEXICO	3.65	-1.81	-1.17	15.1	3.27	-0.57
JAMAICA	1.02	-1.12	-1.21	9.8	1.86	0.43
URUGUAY	6.87	1.71	0.82	3.0	3.44	4.51
CHILE	6.96	2.43	0.26	24.1	-0.75	0.04
TRINIDAD TOBAGO	1.69	5.08	1.34	15.0	1.58	0.66
ARGENTINA	21.86	1.65	1.60	3.8	1.55	1.10
VENEZUELA	4.40	0.51	0.53	7.7	0.74	0.46
LIBYAN ARAB JAMAHIRIYA	13.03	1.20	1.38	8.4	2.36	0.56

SOURCE: FAO.

TABLE VIII : TECHNICAL CONDITIONS OF PRODUCTION IN AGRICULTURE - II

COUNTRY	FERTILIZER PER HECTARE			TRACTORS PER 1000 HECTARES		
	1969 (KG)	GROWTH RATE (%)		1969	GROWTH RATE (%)	
		1969-79	1979-85		1969-79	1979-85
YEMEN ARAB REP	0.4	38.85	0.00	0.34	15.72	2.02
ETHIOPIA	0.4	21.22	0.00	0.19	3.33	0.00
RWANDA	0.3	0.00	39.34	0.09	0.00	0.00
BURKINA FASO	0.2	29.42	11.42	0.02	5.50	3.98
MALI	1.4	9.22	15.01	0.31	2.25	0.00
MALAWI	3.4	10.76	0.00	0.39	2.28	1.68
AFGHANISTAN	2.1	12.35	7.04	0.07	3.78	0.00
LESOTHO	1.2	25.46	0.00	0.96	15.76	2.68
BURUNDI	0.2	6.56	16.30	.	.	0.00
CHAD	0.6	17.93	32.54	0.03	5.19	0.00
NEPAL	2.4	12.03	15.41	0.33	10.40	5.11
MYANMAR	4.0	10.41	14.73	0.43	6.44	2.20
INDONESIA	9.9	13.32	12.15	0.45	0.00	3.39
TANZANIA, UNITED REP.OF	2.3	9.19	0.00	3.72	-0.31	-0.13
CENT.AFR.REP.	1.1	-13.33	0.00	0.03	8.80	2.52
INDIA	8.6	10.70	9.15	0.55	13.45	8.28
BANGLADESH	14.3	12.11	6.78	0.17	7.91	3.85
BENIN	2.9	-18.14	36.75	0.05	0.83	0.00
CHINA	31.5	12.18	5.28	1.76	12.71	3.88
GAMBIA	1.7	23.74	0.00	0.34	-1.58	-2.76
SOMALIA	3.4	-38.17	49.19	0.86	6.12	0.00
TOGO	0.2	21.11	0.00	0.04	11.28	8.67
MADAGASCAR	5.5	-8.71	0.00	1.03	-1.78	0.80
BOTSWANA	1.3	0.00	0.00	1.35	0.91	0.00
UGANDA	0.9	-28.66	.	0.24	4.91	4.07
KENYA	20.9	0.00	11.53	2.97	-1.16	0.93
SUDAN	3.4	0.00	0.00	0.38	7.73	8.13
SIERRA LEONE	1.5	0.00	0.00	0.09	-3.90	18.56
PAKISTAN	16.0	11.57	4.59	0.98	14.59	9.81
GUINEA	1.6	-9.30	-28.42	0.03	10.66	4.85
NIGER	0.1	20.10	0.00	0.01	15.01	0.00
SRI LANKA	44.9	0.00	3.45	6.16	3.14	2.86
CAMEROON	2.4	6.38	8.65	0.02	10.40	11.52
MAURITANIA	0.7	29.08	0.00	0.28	16.32	2.71
THAILAND	7.8	9.51	6.75	0.52	18.90	10.10
NIGERIA	0.3	29.55	15.73	0.07	12.02	3.34
ZAIRE	0.7	8.70	0.00	0.17	5.56	2.82
GUINEA BISSAU	0.4	0.00	.	0.06	10.33	0.65
PHILIPPINES	28.8	4.16	-3.88	0.99	7.24	2.74
JORDAN	7.2	15.26	0.00	7.44	4.35	2.14
EGYPT	122.7	7.04	7.70	5.98	7.41	3.52
SENEGAL	1.6	16.32	0.00	0.06	3.38	1.26
MOZAMBIQUE	2.5	8.85	-37.54	1.33	3.32	0.00

TABLE VIII (CONTD.) : TECHNICAL CONDITIONS OF PRODUCTION IN AGRICULTURE - II

COUNTRY	FERTILIZER PER HECTARE			TRACTORS PER 1000 HECTARES		
	1969 (KG)	GROWTH RATE (%)		1969	GROWTH RATE (%)	
		1969-79	1979-85		1969-79	1979-85
CONGO	10.1	-40.62	61.30	0.96	0.00	0.21
BOLIVIA	1.4	0.00	0.00	0.16	3.62	0.95
GHANA	0.4	29.23	0.00	0.96	2.29	1.05
MOROCCO	12.7	8.45	0.00	1.48	7.99	2.62
PARAGUAY	3.2	-15.11	4.75	2.29	4.17	2.32
MAURITIUS	209.8	0.00	0.00	2.72	1.05	1.08
COTE D'IVOIRE	5.0	6.75	0.00	0.45	4.85	0.63
PAPUA N.GUINEA	5.5	13.06	0.00	3.96	-1.80	-3.44
HONDURAS	14.7	-2.33	0.00	0.93	6.03	0.56
KOREA, REP.OF	231.3	4.76	0.00	0.04	33.87	25.88
ECUADOR	16.3	10.82	0.00	1.14	7.64	4.50
TUNISIA	8.7	3.96	6.89	4.54	0.93	0.00
ZIMBABWE	40.4	0.00	0.00	7.08	1.13	-0.99
SYRIAN ARAB RP	4.9	13.88	11.27	1.49	12.38	9.08
EL SALVADOR	86.1	0.00	0.00	3.83	1.60	0.56
ANGOLA	3.0	0.00	0.00	1.85	4.00	0.07
SWAZILAND	35.3	7.56	0.00	6.58	8.56	6.65
LIBERIA	4.9	11.48	0.00	0.43	5.80	1.27
COLOMBIA	27.9	6.39	2.95	4.39	0.64	0.39
DOMINICAN REP.	26.8	0.00	0.00	1.71	-1.31	0.00
ALGERIA	10.3	0.00	0.00	5.89	0.00	5.17
GABON	0.1	30.93	58.18	2.64	0.00	1.75
GUATEMALA	21.1	11.89	0.00	2.00	1.28	-0.48
GUYANA	29.5	-3.97	0.00	9.05	-2.44	0.54
NICARAGUA	22.5	0.00	0.00	0.37	16.44	1.95
IRAQ	2.9	16.41	0.00	2.40	3.93	10.33
MALAYSIA	39.0	8.34	6.80	1.01	5.65	5.40
IRAN, ISLAMIC REP.OF	5.6	15.68	15.08	1.21	14.94	5.68
FIJI	25.9	9.11	0.00	8.89	6.40	2.47
ZAMBIA	3.3	8.64	0.00	0.55	4.32	-1.03
BRAZIL	11.8	13.35	0.00	2.91	8.88	6.49
PERU	30.0	2.60	-10.81	3.84	0.00	4.59
COSTA RICA	110.6	2.56	0.00	10.20	0.00	0.00
PANAMA	35.3	1.93	-3.59	3.93	5.92	0.59
SAUDI ARABIA	4.8	13.92	37.51	0.69	2.88	6.38
MEXICO	24.2	7.85	6.14	3.77	1.64	5.58
JAMAICA	93.1	-6.30	0.00	6.73	5.00	0.95
URUGUAY	42.1	0.00	-9.97	18.50	1.88	0.59
CHILE	31.2	-5.40	9.19	7.04	-1.28	-0.03
TRINIDAD TOBAGO	82.0	-4.21	-18.66	18.00	0.73	1.53
ARGENTINA	2.4	0.00	0.00	5.29	-0.43	2.69
VENEZUELA	13.4	15.02	0.00	5.06	7.12	2.56
LIBYAN ARAB JAMAHIRIYA	4.3	17.99	10.31	1.83	19.63	3.74

SOURCE: FAO.

TABLE IX : AGRICULTURAL EXPORTS

COUNTRY	AS % OF TOTAL MERCHANDISE EXPORTS			AS % OF AGRICULTURAL GDP		
	1969-71	1978-80	1983-85	1969	1979	1985
YEMEN ARAB REP	55.6	67.5	3.8	1.6	2.2	0.1
ETHIOPIA	80.0	78.0	64.6	11.1	17.9	11.2
RWANDA	60.2	85.0	76.4	6.2	23.2	10.1
BURKINA FASO	73.6	69.0	70.8	12.3	11.2	13.4
MALI	69.5	91.6	91.4	13.2	18.3	32.8
MALAWI	86.5	89.5	91.6	41.8	55.5	65.8
AFGHANISTAN	49.2	47.3	20.1	.	.	.
LESOTHO	75.0	20.6	35.1	38.0	11.4	11.8
BURUNDI	92.6	94.3	88.3	14.5	23.8	17.8
CHAD	99.6	74.6	89.8	26.5	31.9	60.6
NEPAL	92.1	34.0	6.5	11.8	4.1	1.0
MYANMAR	67.0	45.4	54.3	10.8	7.6	3.7
INDONESIA	41.6	13.2	10.4	9.5	14.4	11.3
TANZANIA, UNITED REP.OF	59.4	67.6	67.3	34.7	19.8	8.3
CENT.AFR REP	55.1	43.2	42.5	28.5	11.6	21.8
INDIA	27.2	22.7	21.4	2.2	4.0	3.1
BANGLADESH	35.7	29.4	22.5	6.0	3.2	2.7
BENIN	88.5	67.0	71.5	16.8	8.5	17.3
CHINA	31.8	16.4	13.9	2.6	2.7	5.0
GAMBIA	94.5	70.4	51.8	108.4	47.6	47.3
SOMALIA	75.9	85.4	52.1	15.3	18.6	9.1
TOGO	63.9	31.4	36.3	23.5	30.9	35.2
MADAGASCAR	66.3	66.7	69.5	33.5	24.5	22.4
BOTSWANA	58.0	14.8	8.6	40.6	83.6	94.8
UGANDA	81.1	99.4	91.2	33.6	4.1	.
KENYA	48.2	50.9	60.2	29.8	32.6	38.7
SUDAN	83.5	83.3	72.3	40.7	19.2	18.6
SIERRA LEONE	14.9	31.6	34.3	12.5	24.7	19.9
PAKISTAN	27.8	31.7	24.3	7.9	8.6	8.4
GUINEA	37.4	8.1	6.3	.	4.6	1.9
NIGER	134.2	14.4	19.3	9.3	4.8	8.3
SRI LANKA	90.0	65.0	52.1	54.3	73.9	43.2
CAMEROON	70.2	59.4	44.1	52.1	36.9	29.8
MAURITANIA	19.2	23.2	11.7	33.4	21.2	21.1
THAILAND	65.8	51.0	45.4	24.9	37.8	42.6
NIGERIA	33.4	4.4	3.1	18.2	4.5	1.6
ZAIRE	13.6	16.8	15.8	11.6	5.6	11.5
GUINEA BISSAU	80.0	63.3	67.8	6.3	10.6	8.3
PHILIPPINES	41.6	35.6	25.0	14.6	21.1	11.7
JORDAN	30.5	19.1	10.5	24.9	55.4	23.5
EGYPT	65.7	25.8	18.7	29.0	14.2	5.7
SENEGAL	50.5	32.2	21.4	36.7	35.6	13.0
MOZAMBIQUE	73.6	57.9	47.9	15.7	6.3	.

TABLE IX (CONTD.) : AGRICULTURAL EXPORTS

COUNTRY	AS % OF TOTAL MERCHANDISE EXPORTS			AS % OF AGRICULTURAL GDP		
	1969-71	1978-80	1983-85	1969	1979	1985
CONGO	31.4	4.6	1.1	29.5	10.7	8.5
BOLIVIA	7.8	9.9	4.0	3.0	10.8	2.4
GHANA	75.0	70.1	46.5	27.1	12.1	11.0
MOROCCO	38.2	20.7	11.1	28.1	13.6	9.7
PARAGUAY	57.3	75.2	88.3	16.0	21.8	18.3
MAURITIUS	90.3	67.2	50.7	.	.	.
COTE D'IVOIRE	65.4	67.4	65.5	60.2	70.8	103.6
PAPUA N.GUINEA	67.2	38.3	37.4	27.5	49.5	48.4
HONDURAS	72.8	73.4	68.1	53.6	95.4	69.5
KOREA, REP.OF	6.7	2.1	0.9	2.2	2.1	1.9
ECUADOR	82.5	34.3	18.6	41.4	57.0	26.4
TUNISIA	16.9	9.5	6.0	20.6	19.4	9.3
ZIMBABWE	34.8	35.6	39.1	47.8	76.6	72.1
SYRIAN ARAB RP	65.3	16.5	14.3	35.5	14.9	4.1
EL SALVADOR	65.4	70.6	62.4	50.0	71.3	49.5
ANGOLA	47.3	21.0	8.5	25.7	11.6	.
SWAZILAND	39.8	50.4	46.6	82.6	.	.
LIBERIA	16.6	24.3	27.9	41.2	40.2	32.9
COLOMBIA	77.8	73.7	62.5	27.0	39.1	36.6
DOMINICAN REP.	81.8	55.2	55.6	46.8	47.8	48.8
ALGERIA	16.9	1.4	0.4	41.1	3.8	1.2
GABON	1.5	0.8	0.4	4.0	6.1	2.7
GUATEMALA	62.9	65.4	56.9	.	.	.
GUYANA	39.1	42.3	42.7	.	.	.
NICARAGUA	70.3	82.2	75.9	62.4	126.9	25.2
IRAQ	2.8	0.3	0.3	6.6	2.1	.
MALAYSIA	43.6	32.0	25.2	67.1	73.2	60.6
IRAN ISLAMIC REP.OF	3.9	0.8	0.3	4.5	1.7	.
FIJI	68.7	59.1	46.6	114.7	78.1	54.3
ZAMBIA	0.8	1.1	1.5	3.4	1.0	4.6
BRAZIL	65.5	41.8	32.5	39.1	20.7	35.7
PERU	16.2	11.3	6.7	12.3	21.2	14.8
COSTA RICA	74.9	67.9	60.9	73.9	87.9	81.4
PANAMA	63.4	43.1	48.4	47.0	38.5	33.1
SAUDI ARABIA	0.1	0.0	0.1	0.5	1.4	1.5
MEXICO	47.2	17.1	5.7	19.9	14.0	6.6
JAMAICA	19.2	11.2	13.8	62.6	53.4	68.2
URUGUAY	66.8	40.5	47.9	59.4	36.5	74.7
CHILE	2.8	7.0	9.5	4.9	16.9	28.5
TRINIDAD TOBAGO	6.3	1.9	1.5	84.2	42.4	11.4
ARGENTINA	77.3	66.5	70.5	43.6	46.4	77.2
VENEZUELA	1.0	0.5	0.5	3.6	2.5	2.4
LIBYAN ARAB JAMAHIRIYA	.	.	.	0.6	.	.

SOURCE: FAO.

TABLE X : SHARE OF WAGE EMPLOYMENT IN LABOUR FORCE AND GROWTH OF WAGE EMPLOYMENT IN NON-AGRICULTURE

COUNTRY	WAGE WORKERS AS % OF ALL WORKERS			AVERAGE ANNUAL RATE OF GROWTH OF WAGE EMPLOYMENT (%)	
	1970	1979	1985	1970-79	1979-85
YEMEN ARAB REP	.	.	.	.	.
ETHIOPIA	.	.	.	.	.
RWANDA	.	.	.	.	.
BURKINA FASO	.	.	.	.	.
MALI	.	.	.	.	.
MALAWI	51.71	41.90	35.08	5.80	3.91
AFGHANISTAN	.	.	.	.	.
LESOTHO	.	.	.	.	.
BURUNDI	11.68	12.42	18.91	3.15	11.48
CHAD	.	.	.	.	.
NEPAL	.	.	.	.	.
MYANMAR	.	61.82	60.08	.	2.79
INDONESIA	.	.	.	.	.
TANZANIA, UNITED REP.OF	.	36.70	30.56	.	2.17
CENT.AFR.REP.	.	5.44	3.62	.	-3.51
INDIA	25.68	26.32	24.91	2.68	2.03
BANGLADESH	.	.	.	.	.
BENIN	.	11.53	11.21	.	4.10
CHINA	.	75.27	72.72	.	4.21
GAMBIA	45.71	62.22	42.59	9.82	-2.86
SOMALIA	.	.	.	.	.
TOGO	.	14.48	15.73	.	4.89
MADAGASCAR	.	.	.	.	.
BOTSWANA	81.82	80.68	89.68	10.05	7.95
UGANDA	.	.	.	.	.
KENYA	55.91	56.62	53.22	6.14	4.48
SUDAN	.	.	.	.	.
SIERRA LEONE	.	16.45	13.70	.	0.32
PAKISTAN	.	.	.	.	.
GUINEA	.	.	.	.	.
NIGER	.	14.52	6.45	.	-7.74
SRI LANKA	21.44	23.46	15.05	3.83	-5.04
CAMEROON	20.62	19.59	.	7.36	.
MAURITANIA	.	.	.	.	.
THAILAND	.	.	.	.	.
NIGERIA	.	.	.	.	.
ZAIRE	.	.	.	.	.
GUINEA BISSAU	.	.	.	.	.
PHILIPPINES	.	.	.	.	.
JORDAN	13.54	19.65	20.79	8.71	6.73
EGYPT	.	.	.	.	.
SENEGAL	.	.	.	.	.
MOZAMBIQUE	.	.	.	.	.

TABLE X (CONTD.): SHARE OF WAGE EMPLOYMENT IN LABOUR FORCE AND GROWTH OF WAGE EMPLOYMENT IN NON-AGRICULTURE

COUNTRY	WAGE WORKERS AS % OF ALL WORKERS			AVERAGE ANNUAL RATE OF GROWTH OF WAGE EMPLOYMENT (%)	
	1970	1979	1985	1970-79	1979-85
CONGO	.	.	.	.	.
BOLIVIA	91.14	98.67	79.53	4.17	-0.08
GHANA	.	.	.	.	.
MOROCCO	.	.	.	.	.
PARAGUAY	.	.	.	.	.
MAURITIUS	40.72	62.40	54.71	8.32	1.88
COTE D'IVOIRE	.	32.82	24.76	.	-1.67
PAPUA N.GUINEA	.	.	.	.	.
HONDURAS	.	.	.	.	.
KOREA, REP.OF	.	.	.	.	.
ECUADOR	.	.	.	.	.
TUNISIA	66.45	87.04	78.95	8.19	3.16
ZIMBABWE	.	84.09	77.69	1.75	3.10
SYRIAN ARAB RP	.	.	.	.	.
EL SALVADOR	.	29.05	22.39	.	-0.33
ANGOLA	.	.	.	.	.
SWAZILAND	61.53	75.41	61.73	7.50	1.27
LIBERIA	29.37	51.67	.	10.68	.
COLOMBIA	.	.	.	.	.
DOMINICAN REP.	.	.	.	.	.
ALGERIA	.	.	.	.	.
GABON	.	.	.	.	.
GUATEMALA	.	46.42	38.31	.	0.71
GUYANA	.	.	.	.	.
NICARAGUA	28.67	29.65	.	4.35	.
IRAQ	.	.	.	.	.
MALAYSIA	.	.	.	.	.
IRAN, ISLAMIC REP.OF	.	.	.	.	.
FIJI	64.86	71.03	59.85	5.32	0.49
ZAMBIA	89.80	69.23	50.08	1.16	-0.78
BRAZIL	32.84	.	.	.	.
PERU	.	.	.	.	.
COSTA RICA	.	.	.	.	.
PANAMA	.	.	.	.	.
SAUDI ARABIA	.	.	.	.	.
MEXICO	.	32.23	35.44	.	6.18
JAMAICA	.	.	.	.	.
URUGUAY	.	.	.	.	.
CHILE	.	.	.	.	.
TRINIDAD TOBAGO	.	.	.	.	.
ARGENTINA	.	.	.	.	.
VENEZUELA	.	.	.	.	.
LIBYN ARAB JAMAHIRIYA	.	.	.	.	.

SOURCE:ILO.

TABLE XI : SHARE OF MANUFACTURING AND SERVICES IN TOTAL WAGE EMPLOYMENT IN NON-AGRICULTURE (PERCENTAGES)

COUNTRY	SHARE OF MANUFACTURING			SHARE OF SERVICES		
	1970	1979	1985	1970	1979	1985
YEMEN ARAB REP	..	..	..	..	..	..
ETHIOPIA	..	..	..	..	..	..
RWANDA	..	..	..	..	..	..
BURKINA FASO	..	..	..	..	..	..
MALI	..	..	..	..	..	..
MALAWI	18.4	21.0	27.1	63.0	62.9	63.5
AFGHANISTAN	..	..	..	..	..	..
LESOTHO	..	..	..	..	..	..
BURUNDI	14.0	14.2	14.6	68.9	18.5	57.3
CHAD	..	..	..	..	..	..
NEPAL	..	..	..	..	..	..
MYANMAR	..	22.7	24.5	..	70.4	68.6
INDONESIA	..	..	..	..	..	..
TANZANIA, UNITED REP.OF	..	22.6	17.3	..	61.1	61.6
CENT.AFR.REP.	..	17.9	50.6	..	..	..
INDIA	28.2	28.3	26.5	..	58.6	60.1
BANGLADESH	..	..	..	..	..	..
BENIN	..	10.5	11.1	..	76.0	79.5
CHINA	..	44.3	43.3	..	..	..
GAMBIA	17.4	12.0	5.9	55.7	64.5	59.5
SOMALIA	..	..	..	..	..	..
TOGO	..	13.1	9.3	..	50.1	72.1
MADAGASCAR	..	..	..	..	..	..
BOTSWANA	7.2	7.7	8.7	70.3	65.7	72.7
UGANDA	..	..	..	..	..	..
KENYA	17.9	19.2	17.0	72.3	70.4	75.2
SUDAN	..	..	..	..	..	..
SIERRA LEONE	..	12.3	12.8	..	62.7	59.3
PAKISTAN	..	30.2	..	..	59.2	..
GUINEA	..	..	..	..	..	..
NIGER	..	8.4	11.2	..	26.1	41.4
SRI LANKA	47.2	35.0	39.9	44.2	42.6	51.0
CAMEROON	17.1	34.9	..	..	38.0	..
MAURITANIA	..	..	..	..	..	..
THAILAND	..	..	..	..	..	..
NIGERIA	..	..	..	..	..	..
ZAIRE	..	..	..	..	..	..
GUINEA BISSAU	..	..	..	..	..	..
PHILIPPINES	..	..	..	..	..	..
JORDAN	14.4	11.2	12.9	81.4	77.8	76.4
EGYPT	..	..	..	..	..	..
SENEGAL	..	..	..	..	..	..
MOZAMBIQUE	..	..	..	..	..	..

TABLE XI (CONTD.): SHARE OF MANUFACTURING AND SERVICES IN TOTAL WAGE EMPLOYMENT IN NON-AGRICULTURE (PERCENTAGES)

COUNTRY	SHARE OF MANUFACTURING			SHARE OF SERVICES		
	1970	1979	1985	1970	1979	1985
CONGO	.	.	.	.	.	.
BOLIVIA	16.8	19.3	16.6	63.8	61.0	69.5
GHANA	.	.	.	.	.	.
MOROCCO	.	.	.	.	.	.
PARAGUAY	.	.	.	.	.	.
MAURITIUS	11.1	25.0	38.6	83.8	58.8	53.0
COTE D'IVOIRE	22.1	17.9	24.2	59.4	67.9	74.4
PAPUA N.GUINEA	.	.	.	.	.	.
HONDURAS	.	29.0	31.0	.	60.6	64.0
KOREA, REP.OF	.	.	.	.	.	.
ECUADOR	.	.	.	.	.	.
TUNISIA	.	31.8	.	.	50.1	.
ZIMBABWE	20.6	22.2	21.7	60.4	61.2	64.5
SYRIAN ARAB RP	.	.	.	.	.	.
EL SALVADOR	.	22.9	20.0	.	62.4	70.1
ANGOLA	.	.	21.3	.	.	67.9
SWAZILAND	22.3	19.3	21.2	53.8	54.7	63.9
LIBERIA	7.8	14.2	.	40.2	53.2	.
COLOMBIA	.	.	.	.	.	.
DOMINICAN REP.	.	.	.	.	.	.
ALGERIA	.	.	.	.	.	.
GABON	.	.	.	.	.	.
GUATEMALA	.	20.5	19.5	.	64.0	73.0
GUYANA	.	.	.	.	.	.
NICARAGUA	25.9	23.9	22.1	64.4	67.6	68.4
IRAQ	.	.	.	.	.	.
MALAYSIA	.	.	.	.	.	.
IRAN, ISLAMIC REP.OF	.	.	.	.	.	.
FIJI	19.1	18.2	17.9	58.7	64.8	69.0
ZAMBIA	12.3	13.1	14.8	45.1	54.0	56.0
BRAZIL	.	.	.	.	42.9	.
PERU	.	.	.	.	.	.
COSTA RICA	.	.	.	.	.	.
PANAMA	.	.	.	.	.	.
SAUDI ARABIA	.	.	.	.	.	.
MEXICO	.	44.1	38.6	.	.	.
JAMAICA	.	.	.	.	.	.
URUGUAY	.	.	.	.	.	.
CHILE	.	.	.	.	.	.
TRINIDAD TOBAGO	.	.	.	.	.	.
ARGENTINA	.	.	.	.	.	.
VENEZUELA	.	.	.	.	.	.
LIBYAN ARAB JAMAHIRIYA	.	.	.	.	.	.

SOURCE: ILO.

TABLE XII : AVERAGE ANNUAL RATE OF GROWTH OF WAGE AND TOTAL EMPLOYMENT IN MANUFACTURING (PERCENTAGES)

COUNTRY	WAGE EMPLOYMENT		TOTAL EMPLOYMENT	
	1969-79	1979-85	1969-79	1979-85
YEMEN ARAB REP	.	.	.	.
ETHIOPIA.	.	.	3.96	4.32
RWANDA	.	.	.	.
BURKINA FASO	.	.	1.48	1.34
MALI	.	.	10.64	0.00
MALAWI	7.38	8.36	6.42	0.00
AFGHANISTAN	.	.	9.73	10.71
LESOTHO	.	.	3.82	1.01
BURUNDI	3.40	12.01	3.09	0.00
CHAD	.	.	.	.
NEPAL	.	.	.	.
MYANMAR	.	4.13	.	.
INDONESIA	.	6.76	6.48	6.06
TANZANIA, UNITED REP.OF	.	-3.19	0.00	1.49
CENT.AFR.REP.	.	18.66	0.00	-3.24
INDIA	2.72	0.93	3.81	2.46
BANGLADESH	.	2.47	8.93	4.20
BENIN	.	5.27	0.00	0.00
CHINA	.	3.82	0.00	0.00
GAMBIA	3.36	-13.70	0.00	0.00
SOMALIA	.	.	.	.
TOGO	.	-1.11	0.00	0.00
MADAGASCAR	.	.	2.48	0.00
BOTSWANA	10.99	10.29	10.09	12.64
UGANDA	.	.	.	.
KENYA	7.32	2.32	7.91	3.62
SUDAN	.	.	.	.
SIERRA LEONE	.	0.97	1.83	-2.15
PAKISTAN	.	.	0.96	-0.06
GUINEA	.	.	.	.
NIGER	.	-3.18	0.00	0.00
SRI LANKA	0.00	-2.97	5.54	0.00
CAMEROON	16.23	.	0.00	2.12
MAURITANIA	.	.	.	.
THAILAND	.	.	7.72	-5.28
NIGERIA	.	.	10.82	5.60
ZAIRE	.	.	.	.
GUINEA BISSAU	.	.	.	.
PHILIPPINES	.	-6.76	10.44	6.66
JORDAN	4.97	9.65	6.95	8.23
EGYPT	.	.	3.69	0.00
SENEGAL	.	.	5.12	-5.65
MOZAMBIQUE	.	.	.	.

TABLE XII (CONTD.): AVERAGE ANNUAL RATE OF GROWTH OF WAGE AND TOTAL EMPLOYMENT IN MANUFACTURING (PERCENTAGES)

COUNTRY	WAGE EMPLOYMENT		TOTAL EMPLOYMENT	
	1969-79	1979-85	1969-79	1979-85
CONGO	.	.	.	.
BOLIVIA	5.81	-2.64	2.81	0.00
GHANA	.	.	5.13	-8.04
MOROCCO	.	.	0.00	5.01
PARAGUAY	3.73	.	.	.
MAURITIUS	18.49	9.97	13.75	2.52
COTE D'IVOIRE	.	0.00	7.58	0.00
PAPUA N.GUINEA	.	.	.	.
HONDURAS	.	1.22	.	.
KOREA, REP.OF	12.32	2.90	11.90	1.67
ECUADOR	.	-1.23	9.44	0.00
TUNISIA	.	.	9.48	8.64
ZIMBABWE	2.62	2.68	2.58	4.59
SYRIAN ARAB RP	.	.	5.01	0.00
EL SALVADOR	.	-2.59	0.00	-4.39
ANGOLA	.	.	.	.
SWAZILAND	5.80	2.88	0.00	0.00
LIBERIA	18.17	.	.	0.00
COLOMBIA	.	-2.40	4.62	-1.99
DOMINICAN REP.	.	2.90	0.00	2.66
ALGERIA	.	.	.	.
GABON	.	.	.	.
GUATEMALA	.	-0.12	4.67	-1.23
GUYANA	.	.	.	.
NICARAGUA	3.45	.	6.29	6.45
IRAQ	.	.	.	.
MALAYSIA	.	.	10.39	0.00
IRAN, ISLAMIC REP.OF	.	.	.	.
FIJI	4.81	0.13	.	.
ZAMBIA	1.84	1.27	4.22	1.31
BRAZIL	.	.	8.35	6.17
PERU	3.81	.	0.00	3.68
COSTA RICA	.	.	.	.
PANAMA	.	2.86	2.17	4.81
SAUDI ARABIA	.	.	.	.
MEXICO	.	0.20	2.55	0.00
JAMAICA	.	.	0.00	-3.93
URUGUAY	.	.	4.97	-12.65
CHILE	.	.	-1.84	-9.15
TRINIDAD TOBAGO	.	-1.51	4.93	-8.62
ARGENTINA	.	-4.06	0.00	-3.75
VENEZUELA	.	2.97	9.35	-1.76
LIBYAN ARAB JAMAHIRIYA	12.03	.	0.00	28.10

SOURCES: WAGE EMPLOYMENT - ILO. TOTAL EMPLOYMENT - UNIDO.

TABLE XIII : WAGE LABOUR IN AGRICULTURE

COUNTRY	WAGE-WORKERS AS % OF ALL AGRICULTURAL WORKERS	REGULAR WAGE-WORKERS AS % OF ALL AGRICULTURAL WORKERS		
	C 1980	C 1970	C 1979	C 1985
YEMEN ARAB REP	.	.	.	.
ETHIOPIA	.	.	.	.
RWANDA	2.20	.	.	.
BURKINA FASO	.	.	.	.
MALI	0.50	.	.	.
MALAWI	5.49	2.7	8.2	7.9
AFGHANISTAN	.	.	.	.
LESOTHO	4.10	.	.	.
BURUNDI	1.00	0.6	0.9	0.2
CHAD	.	.	.	.
NEPAL	.	.	.	.
MYANMAR	.	.	.	.
INDONESIA	16.40	.	.	.
TANZANIA, UNITED REP.OF	.	.	1.6	0.8
CENT.AFR.REP.	.	.	.	.
INDIA	37.60	.	0.6	0.6
BANGLADESH	24.90	.	.	.
BENIN	.	.	0.4	0.4
CHINA	.	.	.	.
GAMBIA	.	0.5	0.8	0.3
SOMALIA	.	.	.	.
TOGO	1.00	.	0.3	0.6
MADAGASCAR	.	.	.	.
BOTSWANA	5.40	2.3	1.9	1.5
UGANDA	.	.	.	.
KENYA	.	5.5	4.5	3.6
SUDAN	.	.	.	.
SIERRA LEONE	.	.	0.6	0.6
PAKISTAN	7.50	.	.	.
GUINEA	.	.	.	.
NIGER	.	.	0.0	0.0
SRI LANKA	47.00	22.5	18.3	11.5
CAMEROON	3.30	.	2.6	.
MAURITANIA	.	.	.	.
THAILAND	6.80	.	.	.
NIGERIA	.	.	.	.
ZAIRE	.	.	.	.
GUINEA BISSAU	.	.	.	.
PHILIPPINES	17.10	.	.	.
JORDAN	.	.	.	.
EGYPT	49.90	.	.	.
SENEGAL	.	.	.	.
MOZAMBIQUE	.	.	.	.

TABLE XIII (CONTD.): WAGE LABOUR IN AGRICULTURE

COUNTRY	WAGE-WORKERS AS % OF ALL AGRICULTURAL WORKERS	REGULAR WAGE-WORKERS AS % OF ALL AGRICULTURAL WORKERS		
	C 1980	C 1970	C 1979	C 1985
CONGO	.	.	.	.
BOLIVIA	12.30	.	.	.
GHANA	.	.	.	.
MOROCCO	.	.	.	.
PARAGUAY	15.10	.	.	.
MAURITIUS	.	68.9	65.1	53.5
COTE D'IVOIRE	.	.	3.0	3.1
PAPUA N.GUINEA	.	.	.	.
HONDURAS	29.50	.	.	.
KOREA, REP.OF	11.80	.	.	.
ECUADOR	28.9	.	.	.
TUNISIA	27.6	.	.	.
ZIMBABWE	.	17.1	15.8	.
SYRIAN ARAB REP	.	.	.	.
EL SALVADOR	56.0	.	.	.
ANGOLA	.	.	.	3.2
SWAZILAND	.	11.4	15.5	11.9
LIBERIA	.	1.9	6.3	.
COLOMBIA	.	.	.	.
DOMINICAN REP.	19.4	.	.	.
ALGERIA	.	.	.	.
GABON	.	.	.	.
GUATEMALA	32.0	.	33.9	19.1
GUYANA	.	.	.	.
NICARAGUA	.	0.3	1.0	7.6
IRAQ	15.5	.	.	.
MALAYSIA	31.8	.	.	.
IRAN, ISLAMIC REP.OF	18.3	.	.	.
FIJI	19.2	4.7	2.4	2.6
ZAMBIA	.	3.0	2.3	2.2
BRAZIL	38.4	.	.	.
PERU	20.2	.	.	.
COSTA RICA	63.4	.	.	.
PANAMA	27.7	.	.	.
SAUDI ARABIA	.	.	.	.
MEXICO	23.1	.	.	.
JAMAICA	.	.	.	.
URUGUAY	50.8	.	.	.
CHILE	61.3	.	.	.
TRINIDAD TOBAGO	.	.	.	.
ARGENTINA	52.9	.	.	.
VENEZUELA	39.6	.	.	.
LIBYAN ARAB JAMAHIRIYA	.	.	.	.

SOURCE: ILO.

TABLE XIV : REAL WAGE RATES IN AGRICULTURE AND MANUFACTURING: AVERAGE ANNUAL RATE OF GROWTH (PERCENTAGES)

COUNTRY	AGRICULTURE		MANUFACTURING	
	1969-79	1979-85	1969-79	1979-85
YEMEN ARAB REP	.	.	.	.
ETHIOPIA	.	.	.	.
RWANDA	.	.	.	.
BURKINA FASO	.	.	.	.
MALI	.	.	.	.
MALAWI	-3.39	-0.71	-1.94	-4.32
AFGHANISTAN	.	.	.	.
LESOTHO	.	.	.	.
BURUNDI	.	.	-6.09	13.51
CHAD	.	.	.	.
NEPAL	.	.	.	.
MYANMAR	-11.38	-3.32	-8.31	-0.35
INDONESIA	0.00	.	.	.
TANZANIA, UNITED REP.OF	.	.	-2.65	.
CENT.AFR.REP.	.	.	.	.
INDIA	0.00	2.99	2.97	.
BANGLADESH	-2.00	-0.62	.	-2.57
BENIN	.	.	.	.
CHINA	.	.	-0.64	3.38
GAMBIA	.	.	.	.
SOMALIA	.	.	.	.
TOGO	.	.	.	.
MADAGASCAR	.	.	.	.
BOTSWANA	.	.	.	.
UGANDA	.	.	.	.
KENYA	1.38	-5.46	-4.26	-5.33
SUDAN	.	.	.	.
SIERRA LEONE	.	.	-4.02	.
PAKISTAN	3.93	16.49	.	.
GUINEA	.	.	.	.
NIGER	.	.	.	.
SRI LANKA	6.52	-3.39	5.79	-2.16
CAMEROON	.	.	.	.
MAURITANIA	.	.	.	.
THAILAND	.	.	.	.
NIGERIA	.	.	.	.
ZAIRE	.	.	.	.
GUINEA BISÇAO	.	.	.	.
PHILIPPINES	.	-7.69	.	.
JORDAN	.	.	.	0.36
EGYPT	.	-7.66	2.17	.
SENEGAL	.	.	.	.
MOZAMBIQUE	.	.	.	.

TABLE XIV (CONTD.) : REAL WAGE RATES IN AGRICULTURE AND MANUFACTURING: AVERAGE ANNUAL RATE OF GROWTH (PERCENTAGES)

COUNTRY	AGRICULTURE		MANUFACTURING	
	1969-79	1979-85	1969-79	1979-85
CONGO	.	.	.	.
BOLIVIA	.	.	1.13	-13.44
GHANA	.	-6.91	.	.
MOROCCO	.	.	.	.
PARAGUAY	.	.	.	.
MAURITIUS	4.21	.	0.73	.
COTE D'IVOIRE	.	.	.	.
PAPUA N.GUINEA	.	.	6.33	.
HONDURAS	.	.	4.75	-9.35
KOREA, REP.OF	8.63	2.61	9.49	2.39
ECUADOR	.	.	.	-6.68
TUNISIA	.	.	.	.
ZIMBABWE	3.04	8.21	2.42	2.04
SYRIAN ARAB RP	.	.	.	.
EL SALVADOR	.	.	.	-3.67
ANGOLA	.	.	.	.
SWAZILAND	.	.	3.98	-13.58
LIBERIA	.	.	.	.
COLOMBIA	.	-1.16	.	.
DOMINICAN REP.	.	.	.	.
ALGERIA	.	.	.	.
GABON	.	.	.	.
GUATEMALA	-3.33	.	.	6.67
GUYANA	-3.14	.	0.72	-17.48
NICARAGUA	-4.41	.	-2.01	.
IRAQ	.	.	.	.
MALAYSIA	1.74	.	.	.
IRAN, ISLAMIC REP.OF	.	.	.	.
FIJI	5.19	-8.49	4.73	-1.27
ZAMBIA	.	.	.	.
BRAZIL	5.19	.	.	.
PERU	.	.	-4.32	-3.12
COSTA RICA	5.49	-4.22	3.82	-6.20
PANAMA	-0.64	2.04	-0.61	.
SAUDI ARABIA	.	.	.	.
MEXICO	0.08	-4.07	-0.45	-5.16
JAMAICA	.	.	.	.
URUGUAY	-3.53	-6.48	.	.
CHILE	.	-1.19	.	14.82
TRINIDAD TOBAGO	.	.	.	.
ARGENTINA	-6.01	.	7.10	-45.87
VENEZUELA	.	.	.	.
LIBYAN ARAB JAMAHIRIYA	.	.	.	.

SOURCES: ILO; RAY(1987); DE JANVRY, SADOULET AND WILCOX(1986).

TABLE XV : RURAL POVERTY, OUTPUT PER WORKER AND LAND DISTRIBUTION

COUNTRY	% OF RURAL POPULATION IN POVERTY	AGRICULTURAL OUTPUT PER AGRICULTURAL WORKERS CONS.1970 US$	GINI COEFFICIENTS OF LAND DISTRIBUTION	
	C 1977	1976/78	C 1970	C 1980
YEMEN ARAB REP	.	172.6	.	0.64
ETHIOPIA	65.0	74.7	.	0.43
RWANDA	90.0	59.9	.	.
BURKINA FASO	.	55.0	.	.
MALI	48.0	148.7	.	.
MALAWI	85.0	84.1	0.36	0.36
AFGHANISTAN	36.0	.	.	.
LESOTHO	55.0	74.5	0.39	.
BURUNDI	85.0	80.5	.	.
CHAD	56.0	105.0	0.38	.
NEPAL	61.0	116.6	.	0.60
MYANMAR	40.0	163.8	.	.
INDONESIA	51.0	149.6	0.62	.
TANZANIA, UNITED REP.OF	60.0	87.3	.	.
CENT.AFR.REP.	.	93.4	0.37	.
INDIA	50.7	136.2	0.64	0.62
BANGLADESH	86.0	220.5	.	0.42
BENIN	65.0	101.0	.	.
CHINA	.	102.1	.	.
GAMBIA	40.0	114.5	.	.
SOMALIA	70.0	163.2	.	.
TOGO	.	130.9	.	0.45
MADAGASCAR	50.0	114.4	.	0.80
BOTSWANA	55.0	240.9	.	.
UGANDA	.	180.9	.	.
KENYA	55.0	185.4	.	0.77
SUDAN	85.0	266.1	.	.
SIERRA LEONE	65.0	180.2	0.44	.
PAKISTAN	29.0	288.9	0.52	0.54
GUINEA	.	203.4	.	.
NIGER	35.0	129.0	.	0.32
SRI LANKA	.	270.5	.	0.62
CAMEROON	40.0	196.2	0.44	.
MAURITANIA	.	166.4	.	0.58
THAILAND	34.0	188.6	.	0.45
NIGERIA	.	194.5	.	.
ZAIRE	80.0	140.5	0.37	.
GUINEA BISAO	.	195.7	.	.
PHILIPPINES	41.0	347.1	0.51	0.53
JORDAN	17.0	771.1	.	0.69
EGYPT	25.0	511.2	.	.
SENEGAL	.	138.4	.	.
MOZAMBIQUE	.	139.1	.	.

TABLE XV (CONTD.): RURAL POVERTY, OUTPUT PER WORKER AND LAND DISTRIBUTION

COUNTRY	% OF RURAL POPULATION IN POVERTY	AGRICULTURAL OUTPUT PER AGRICULTURAL WORKERS CONS.1970 US$	GINI COEFFICIENTS OF LAND DISTRIBUTION	
	C 1977	1976/78	C 1970	C 1980
CONGO	.	153.2	0.29	.
BOLIVIA	85.0	367.4	.	.
GHANA	.	541.2	.	.
MOROCCO	45.0	428.1	.	.
PARAGUAY	50.0	703.9	.	0.94
MAURITIUS	12.0	733.6	.	.
COTE D'IVOIRE	.	253.9	0.42	.
PAPUA N.GUINEA	75.0	236.5	.	.
HONDURAS	54.9	449.1	0.78	.
KOREA, REP.OF	11.0	687.0	0.37	0.30
ECUADOR	65.0	574.0	.	.
TUNISIA	15.0	625.5	.	.
ZIMBABWE	.	127.0	.	.
SYRIAN ARAB RP	.	990.5	.	.
EL SALVADOR	32.0	637.2	0.81	.
ANGOLA	.	179.8	.	.
SWAZILAND	50.0	236.8	.	.
LIBERIA	.	273.1	0.73	.
COLOMBIA	.	947.7	0.86	.
DOMINICAN REP	43.0	714.8	0.79	.
ALGERIA	.	494.1	0.72	.
GABON	.	130.9	0.47	.
GUATEMALA	25.0	.	.	0.85
GUYANA	.	908.4	.	.
NICARAGUA	19.0	700.5	.	.
IRAQ	40.0	478.6	0.65	.
MALAYSIA	37.7	854.6	.	.
IRAN, ISLAMIC REP.OF	.	468.8	.	.
FIJI	30.0	749.4	.	0.84
ZAMBIA	.	249.7	.	.
BRAZIL	.	744.8	0.84	0.86
PERU	.	671.3	.	.
COSTA RICA	.	1328.8	0.83	.
PANAMA	30.0	786.4	0.78	0.84
SAUDI ARABIA	.	85.7	0.79	0.83
MEXICO	.	739.1	.	.
JAMAICA	80.0	421.5	0.80	0.82
URUGUAY	.	1520.4	0.82	0.84
CHILE	.	1114.9	.	.
TRINIDAD TOBAGO	39.0	811.2	.	.
ARGENTINA	.	2098.5	0.87	.
VENEZUELA	.	1210.4	0.92	.
LIBYAN ARAB JAMAHIRIYA	.	516.7	.	.

SOURCE: RURAL POVERTY - WORLD BANK GINI COEFFICIENTS - FAO.